The Chemistry of the XXI Century

MOLECULAR MODELING

The Chemistry of the XXI Century

MOLECULAR MODELING

Rio de Janeiro, Brazil   May 25 – 27, 1992

Editor

Marco Antonio Chaer Nascimento

Departamento de Fisico-Química
Universidade Federal do Rio de Janeiro, Brazil

World Scientific
Singapore • New Jersey • London • Hong Kong

₀6075873

CHEMISTRY

Published by

World Scientific Publishing Co. Pte. Ltd.
P O Box 128, Farrer Road, Singapore 9128
USA office: Suite 1B, 1060 Main Street, River Edge, NJ 07661
UK office: 73 Lynton Mead, Totteridge, London N20 8DH

MOLECULAR MODELING

ISBN 981-02-1620-3

Printed in Singapore.

PREFACE

The object of molecular modeling is the generation, manipulation and representation of **realistic** three-dimensional molecular structures, or of a system of interacting (like or unlike) molecules, with the purpose of understanding the physicochemical properties and macroscopic phenomena at the molecular level.

Ab-initio quantum chemistry calculations have been extensively used to characterize the structure and properties of a large variety of molecules. However, in spite of the recent progress in the methodologies, ab-initio calculations remain far too slow to be used in realistic modeling of important systems such as catalysts, biomolecules, polymers, ceramics and other materials. Nevertheless, ab-initio calculations will remain extremely useful in the determination of force field parameters obtained from calculations on representative sub-units of those complex systems. These force fields, derived from ab-initio calculations, can then be used to model systems, far beyond the reach of ab-initio quantum chemistry, by molecular mechanics and dynamics methods.

In the last ten years, as a result of the advent of powerful affordable personal computers together with interactive, high-resolution computer graphics terminals and the development of very sophisticated multifunctional software packages, molecular modeling has found widespread usage in different research areas.

Since the activity of a drug, or the specificity of a catalyst or the mechanical properties of a polymer are direct consequences of their chemical structures, molecular modeling provides a natural **unification** of such apparently macroscopically-unrelated areas of research.

Molecular modeling has become an important tool in everyday research and its usefulness cannot be ignored by anyone conducting research in medical, natural and engineering sciences. This book is an attempt to show molecular modeling as a new multidisciplinary area of research that transcends the boundaries traditionally separating biology, chemistry and physics. To this purpose, leading scientists were invited to present applications of molecular modeling to a variety of important problems such as drug design, materials design, protein modeling, catalyst modeling, mechanical properties of materials and properties of glass. The book contains part of the material presented in the meeting on Molecular Modeling, held in Rio de Janeiro (May 25–27, 1992) and, as will become evident in reading, molecular modeling has already proven itself extremely useful to the scientific community. Although its importance should not be exaggerated, molecular modeling will certainly be one of the most important research tools in the next century.

The success of a meeting depends on a series of different factors and proper financial support is one of the most important. This meeting would not have been possible without the support of IBM Brasil. I would like to express my gratitude to José Paulo Schiffini, of IBM Brasil, for his support, encouragement and enthusiasm. Financial support from Biosym Technologies Inc. is also gratefully acknowledged.

The contribution of many others was essential to the realization of the meeting, but I am particularly indebted to one of my students, João Otávio Milan de Albuquerque Lins, and to Angela Maria Lopes, Maria Celeste Pereira and Paulo Jorge da Costa, members of the Physical Chemistry Department staff for their professionalism and dedication in dealing with the meeting organization.

Marco Antonio Chaer Nascimento
Rio de Janeiro
July, 1993

CONTENTS

CONTENTS

ATOMISTIC MODELLING OF ZEOLITIC MATERIALS

R.A. van Santen and A.J.M. de Man
Schuit Institute of Catalysis
Eindhoven University of Technology
Eindhoven, The Netherlands

1. Introduction.

Zeolites and zeolitic materials form a class of crystalline substances with an increasing technological importance and are therefore subject to various modelling methods assisting the experimental research on their structures and properties. Zeolites are, strictly spoken, crystalline alumino–silicates with an open framework structure containing pores and cavities on a molecular scale[1]. Some zeolites occur in nature, some only exist in a synthesized form[2]. The atomic structure of Faujasite, a typical zeolite, is given in Figure 1. The pores and cages are so large that they can absorb molecules, even large hydrocarbons. The molecules adsorbed in the zeolite cages and pores can move rather unhindered in and out the crystals.

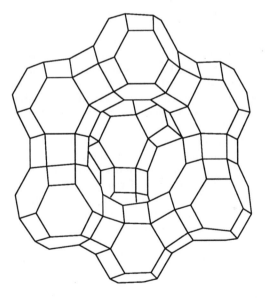

Figure 1 Schematic drawing of the faujasite structure. Only faces spanned by lines between silicon (or aluminum) atoms are drawn. Oxygen atoms are located near the middle of each line.

The silicon or aluminum atoms in the framework can be substituted by a large variety of other elements, such as germanium, titanium, gallium, beryllium. All these atoms, which are tetrahedrally surrounded by oxygen, are generally called the T–atoms. The general framework composition is TO_2. When aluminum (or any other trivalent atom) is present in the zeolite the framework by itself is not electrically neutral any more and charge compensating cations like sodium and potassium have to be introduced inside the zeolite pores and cages. These cations can be exchanged rather easily. This ion exchange capacity is one of the profitable features of zeolites, for instance for use in detergents.

When extra–framework cations are replaced by protons zeolites become acidic. This, in combination with the open, well defined structure, makes zeolites suitable as shape selective solid acid catalysts. Whereas the ion exchange capacity is controlled by composition rather than zeolite structure, the use as a catalyst is strongly structure dependent. Therefore the synthesis of zeolites with new structures is one of the main topics in zeolite research. Closely related to zeolites are alumino–phosphate molecular sieves[3] ($AlPO_4$'s), which can have topologies similar and sometimes equal to those of zeolites, and clathrasils[4-7], silica structures that trap guest molecules during their synthesis. Beside their importance as model systems for zeolites clathrasils are also interesting as non-linear optical materials[8]. The T–atoms of $AlPO_4$'s can be substituted by a large variety of elements[9]

Because zeolite synthesis is time consuming and mainly based on trial and error the need for theoretical predictions of zeolite properties is evident. A large intensification of the theoretical study of zeolites can be seen in the last ten years[10-39]. Most of the early atomistic modelling studies on zeolites were related to their practical use as a catalyst, molecular sieve or ion exchange resin. In other words, these studies were mainly focussed on the relation between the zeolite framework and the molecules or ions present within the zeolite. For the interaction between adsorbed molecules and framework oxygen atoms most studies use a combination of a Coulomb and a 6–12 Lennard–Jones[40-42], Buckingham[39,43] or Morse[44,45] potential while for the interaction between molecules and T–atoms usually only a Coulomb term has been applied. Because the guest–host interaction potentials are strongly ionic a large number of studies were based on purely electrostatic calculations[46-51]. The framework itself was usually treated in a simplified way: the framework atoms are kept fixed on positions derived from X–ray diffraction (X.R.D.) or neutron diffraction (N.D.) experiments.

Here we will focus on modelling the structure and properties of the bare zeolitic framework. The experimental derivation of the structure by conventional diffraction methods is very difficult because the crystallites of synthetic zeolites are usually very small[52], enabling powder diffraction methods only, while the number of atoms per unit cell might be large. Structure refinement based on experimental data has been assisted extensively by the D.L.S. method[53], in which a structure with a given symmetry and unit cell dimensions is changed in such a way that the T−O distances match a prescribed physical value. We will show in section 5 that a combination of experimental data and atomistic modelling using various interatomic potentials is more appropriate for zeolite structure refinement. Properties that can be calculated are lattice energies, free energies, optimized atomic positions and unit cell vectors, dielectric and elastic constants, vibrational spectra and crystallite shapes[54].

2. Modelling methods.

The first step in modelling framework properties of a zeolite is the derivation of the atomic structure. As pointed out in the previous section structures solely based on experimental data are usually not good enough for our purposes. One can obtain a better structure by minimizing its lattice energy with respect to the atomic positions and the unit cell dimensions[36,55]. The lattice energy is calculated with the THBREL program[33,34,56], using two–body and three–body interatomic potentials. Some of these potentials cannot reproduce the absolute value of the lattice energy, while they are very well able to refine structures. Absolute values of lattice energies are currently of limited use because for only a few zeolites experimental values have been derived[57,58]. Because we do not perform free energy calculations the results essentially apply to properties at a temperature of zero Kelvin. Some nice examples of the use of free energy minimizations have been given by Parker[59]. The energy minimization should be done with care because the experimentally determined symmetry can be too high and the minimization procedure might preserve some invalid symmetry elements[37,39].

Initially the zeolitic frameworks are taken to be neutral because the introduction of aluminum in a zeolite causes several problems in modelling. First of all, in most structure refinements no explicit aluminum atom positions are given, but merely positions of T–atoms that are partially occupied by aluminum. Modelling requires fully occupied crystallographic sites. One usually cannot simply lower the symmetry of the crystal in order to obtain fully occupied aluminum sites from the original partially occupied positions because very low symmetries would be needed and it is not even sure whether the aluminum sites are really ordered. The generation of aluminum positions is somewhat facilitated by the Löwenstein rule[60,61] which states that two neighboring T–sites cannot both be occupied by aluminum atoms. The extra–framework cations needed to compensate the charge make calculations even more difficult because their positions are usually even more uncertain than the aluminum coordinates, due to both static and dynamic disorder.

Another difficulty in modelling zeolites is the derivation of potential parameters once a functional form for the interatomic potentials is choosen. A derivation based on experimental data, such as structures, elastic constants and vibrational spectra, requires well defined model systems[35,62−65]. Such systems are rare. For instance, the fitting of potential parameters on vibrational spectra requires a reliable group theoretical assignment of the vibrational modes, which is virtually absent for zeolites. Even the group theoretical assignment of vibrational modes of natural sodalite[66], one of the simplest zeolitic structures, is still under discussion[67]. Potential parameters derived from *ab–initio* quantum chemical calculations on molecular substructures may provide the solution for the potential parameter problem[12,20,68−78] but one should carefully check if the influence of the surrounding zeolite on the substructure is properly accounted for[79]. Therefore large molecules may be needed, thus increasing the costs of the quantum–chemical calculation. In such cases it might be feasible to perform *ab–initio* quantum chemical calculations on periodic systems. Recently Hartree–Fock[80,81] and density functional[82,83] programs have become available that can handle crystallographic unit cells of the size of small zeolites. Since full geometry optimizations are still somewhat problematic with these programs one usually performs quantum–chemical calculations on some well known structures or systems of which only a few structural parameters have to be optimized[84]. Two– and three–body atomic potentials can than be derived from the energy hypersurface of these optimized structures[85]. Another use of quantum–chemical calculations in deriving force fields is the Hessian–Biased approach[86], in which quantum–chemically calculated vibrational modes and experimental vibrational frequencies are used. In general quantum–chemically calculated vibrational frequencies are not accurate enough, but the (experimentally inaccessible) calculated vibrational modes are reasonably accurate. The method has been used for polymers[87] but not yet for zeolitic systems.

3. Potential development.

Most interatomic potentials used in modelling zeolite frameworks consist of a long range Coulomb term and a short range term. A potential that has been used widely is the silica rigid ion model of Catlow *et al.*[36], which describes Si–O and O–O short range interactions by a Buckingham potential and also comprises harmonic O–Si–O bond bending terms. The three–body bending term is harmonic and only calculated for Si–O pairs with the same silicon atom (*i.e.* for the bending of a tetrahedral angle). The complete potential, which we shall simply refer to as the Rigid Ion model, is represented in Eq. 1.

$$V = \frac{1}{2} \sum_i \left(\sum_{j \neq i} \frac{q_i q_j}{4\pi \epsilon_0 r_{ij}} + {\sum_{j \neq i}}' \left(A_{ij} e^{-r_{ij}/\rho_{ij}} - \frac{C_{ij}}{r_{ij}^6} \right) + {\sum_{j \neq i, l \neq \{i,j\}}}'' k_{jil} \left(\theta_{jil} - \theta_{jil}^0 \right)^2 \right) \quad (1)$$

$$
\begin{array}{lll}
\text{with} & V & = \text{potential energy} \\
& q_i & = \text{charge of atom } i \\
& r_{ij} & = \text{distance between atoms } i \text{ and } j \\
& A_{ij}, \rho_{ij}, C_{ij} & = \text{parameters of the Buckingham term,} \\
& & \quad \text{depending on the atom types of } i \text{ and } j \\
& \theta_{jil} & = \text{angle between bonds } ij \text{ and } il \\
& k_{jil} & = \text{three–body force constant} \\
& \theta_{jil}^0 & = \text{equilibrium angle between bonds } ij \text{ and } il \\
& {\sum}'_{j \neq i} & = \text{sum over atoms } j \text{ within of 10 Å from atom } i \\
& {\sum}''_{j \neq i, l \neq \{i,j\}} & = \text{double sum over atoms } j \text{ and } l \text{ neighboring } i
\end{array}
$$

A second model for silica, the Shell model[35,88,89], is an extension of the Rigid Ion model. In this model atomic polarizabilities are taken into account. Atoms are thought of as consisting of a charged point mass (the core) and a massless, isotropic shell with charge q^{shell} that are connected by a harmonic "spring" with force constant K. The polarizability α of a free ion described in this way is given in Eq. 2.

$$\alpha = \frac{(q^{shell})^2}{4\pi\epsilon_o K} \qquad (2)$$

In general only oxygen ions are described as a core–shell system because the electronic polarizability of a free oxygen ion is significantly larger than that of any T–atom[90]. For atoms that are described as a core–shell system short range interactions with other atoms are calculated *via* the shell. It should be noted that here the shell can be regarded as a point charge. In other models, like the modified polarizable ion model of Iishi[91], spherical Gaussian charge distributions are used. No and Jhon *et al.*[47,65] used explicit atomic polarizabilities and calculated the electric field at each atom position to account for polarization effects. Rigid Ion and Shell model Si–O potential parameters were obtained[36,88] by modelling the structure and elastic constants of α–quartz. The O–O short range interaction was derived[92] from a fit on the potential energy curve of $O^- - O^-$ that was obtained from Hartree–Fock calculations. The short range terms are calculated for all atoms within a range of 10 Å of each atom. Coulomb terms are calculated with the Ewald summation technique[36,93].

A more recent potential for silica has been derived by van Beest *et al.*[68]. It is a two–body rigid ion potential with a Coulomb term using partial charges and Buckingham short range interactions. The values of the parameters were derived from a fit on potential energy surfaces obtained from *ab–initio* quantum–mechanical data for $Si(OH)_4$ and $(OH)_3SiOSi(OH)_3$ while the fit was constrained by some crystallographic data on α–quartz. This potential set will be referred to as the Partial Charge model. The same authors also derived a partial charge potential set for $AlPO_4$ that is consistent with the silica model. The parameter values for all these potentials are given in Table 1.

The Rigid Ion model and the Shell model were mainly developed for structure refinement[31,36,88] and not for vibrational spectra, one of the most frequently used characterization tools in zeolite chemistry[94–104]. Tests of these potentials show two serious discrepancies between experimental and calculated infrared and Raman spectra of α–quartz[62]: the calculated frequencies of symmetric and asymmetric stretching modes (between 700 and 1400 cm^{-1}) are too small and the splitting between the longitudinal and transversal optical modes (the TO–LO splitting) is too large. Several calculated properties of α–quartz are shown in Table 2, in which the TO–LO splitting for a certain wave vector $\vec{k}$ is expressed by means of the

Table 1. Potential parameters[36,68,88].

Parameter	Rigid ion	Shell	Partial Charge	
q_{Si}	+4.0	+4.0	+2.4	e
q_{Al}	—	—	+1.4	e
q_P	—	—	+3.4	e
q_O^{core}	−2.0	+0.86902	−1.2	e
q_O^{shell}	—	−2.86902	—	e
$K_{core-shell}$	—	74.92	—	eV/Å²
A_{SiO}	1584.167	1283.907	18003.757	eV
A_{ALO}	—	—	16008.535	eV
A_{PO}	—	—	9034.208	eV
A_{OO}	22764.3	22764.3	1388.773	eV
ρ_{SiO}	0.32962	0.32052	0.20521	Å
ρ_{AlO}	—	—	0.20848	Å
ρ_{PO}	—	—	0.19264	Å
ρ_{OO}	0.149	0.149	0.36232	Å
C_{SiO}	52.645	10.6616	13.3538	eVÅ⁶
C_{AlO}	—	—	130.5659	eVÅ⁶
C_{PO}	—	—	19.8793	eVÅ⁶
C_{OO}	27.88	27.88	175.00	eVÅ⁶
θ^0	109.47	109.47	—	°
k	4.5815	2.09724	—	eV/rad²

plasmon frequency $\omega_{pl,\vec{k}}$ given in Eq. 3.

$$\left(\omega_{pl,\vec{k}}\right)^2 = \frac{1}{N_{\vec{k}}} \sum_{modes} \left(\left(\omega_{i,\vec{k}}^{LO}\right)^2 - \left(\omega_i^{TO}\right)^2\right) \tag{3}$$

The frequencies ω_i^{TO} of the transversal modes can be taken from a calculation at $\vec{k} = \vec{0}$. The frequencies $\omega_{i,\vec{k}}^{LO}$ depend on the wave vector $\vec{k}$, as does the number of shifted frequencies $N_{\vec{k}}$. In Table 2 values of $\omega_{pl,\vec{k}}$ are given for $\vec{k} = (0.001, 0, 0)$ and $\vec{k} = (0, 0, 0.001)$ (reciprocal lattice units). The first vector splits the E modes of

α–quartz and the second one shifts the A_2 modes. The vectors are close enough to $\vec{k} = \vec{0}$ to avoid dispersion effects, so the values of $\omega_{\mathrm{pl},\vec{k}}$ can be regarded as limits for $\vec{k} \to \vec{0}$. The plasmon frequency can be related[105] to an effective ionic charge Z'_{eff} by regarding the crystal as consisting of a set of oscillators with reduced mass μ and volume v_a (Eq. 4).

$$\left(\omega_{\mathrm{pl},\vec{k}}\right)^2 = \frac{4\pi\left(Z'_{\mathrm{eff}}\right)^2}{\mu v_a \epsilon_{\vec{k}}(\infty)} \tag{4}$$

The high frequency dielectric constant $\epsilon_{\vec{k}}(\infty)$ used in Eq. 4 can be derived from experiments.

From the overestimation of the TO–LO splitting in the Rigid Ion and Shell model one can conclude that the charge of the ions should be lowered and, consequently, that the short range interaction should be increased. This was the basis for the development of the Partial Charge model, which indeed shows a better agreement between calculated and experimental stretching frequencies and is also better in predicting other properties of α–quartz, but does not perform very well for the low frequency modes when compared with the Shell model. This is partly due to the omission of three–body terms in the Partial Charge model. Apart from the ionic models one can also use harmonic Generalized Valence Force Fields[108] (GVFFs) for simulating spectra. For pure silica systems we have used the GVFF derived by Etchepare[109] by fitting nearest neighbor stretching, bending and torsion forces (and some cross terms) on the spectra of α–quartz and β–quartz. For AlPO$_4$'s a GVFF has been developed[110] with the same method using spectra of α–berlinite. Parameter values of both GVFF's are given in Table 3. Besides force fields derived from spectral and structural data there exists a large variety of silica force fields based on empirical "rules of thumb"[101,111–113] These force fields possess only a limited number of interaction terms. However, they can be very useful in providing starting values for the fitting procedure of more intricate force fields. Calculated and experimental spectra of α–quartz are shown in Figure 2. The GVFF reproduces the experimental spectra significantly better than the ionic potentials but cannot be used for structure relaxation, so in practice both kinds of models are applied.

Table 2.   Comparison of potentials for α-quartz.

	Experiment[106]	Rigid ion	Shell	Partial Charge
Unit cell vectors (Å)				
a	4.9130	5.1051	4.8366	4.9384
c	5.4046	5.6312	5.3469	5.4489
Elastic constants (10^{11} dyn/cm^2)				
C_{11}	8.6832	20.2343	9.4707	9.0558
C_{33}	10.598	21.2907	11.6080	10.7068
C_{12}	0.709	6.6353	1.1837	0.8118
C_{13}	1.193	8.6264	1.9683	1.5250
C_{14}	−1.8064	0[1]	−1.4518	−1.7661
C_{44}	5.8257	6.3078	5.0056	5.0276
C_{66}	3.9871	6.7995	3.8170	4.1220
K_0[2]	3.745	12.11	4.459	4.011
Static dielectric constants				
ϵ_{11}^0	4.514	3.8993	4.7396	1.9537
ϵ_{33}^0	4.640	4.0689	5.0136	1.9948
High frequency dielectric constants				
ϵ_{11}^∞	2.356	[3]	2.1163	[3]
ϵ_{33}^∞	2.383	[3]	2.1397	[3]
Plasmon frequencies (cm^{-1})				
$\omega_{\mathrm{pl},x}$	139.57	557.80	366.80	351.53
$\omega_{\mathrm{pl},z}$	98.55	394.43	254.24	248.56

[1] The Rigid Ion model relaxation of α-quartz results in a structure resembling β-quartz. Experimental values of the elastic constants for β-quartz are[107]: C_{11}=11.84, C_{33}=10.70, C_{12}=1.90, C_{13}=3.20, C_{44}=3.585 and C_{66}=4.997 10^{11} dyn/cm^2 (at 600°C).

[2] K_0 is the bulk modulus.

[3] Rigid ion models do not give high frequency dielectric constants.

4. Vibrational spectra of zeolites.

The use of vibrational spectroscopy for structural characterization of zeolites is not straightforward. One of the most important topics in this field is the question

Table 3. Parameters of the Generalized Valence Force Fields.

$SiO_2^{(109)}$		$AlPO_4^{(110)}$		
Bond stretching				
K_{Si-O}	5.943	K_{Al-O}	4.197	mdyn/Å
		K_{P-O}	5.145	mdyn/Å
Bond bending				
K_{O-Si-O}	0.729	K_{O-Al-O}	0.412	mdynÅ/rad^2
		K_{O-P-O}	0.998	mdynÅ/rad^2
$K_{Si-O-Si}$	0.126	K_{Al-O-P}	0.197	mdynÅ/rad^2
Torsion				
$K_{Si-O-Si-O}$	0.0056	$K_{P-O-Al-O}$	[1]	mdynÅ/rad^2
		$K_{Al-O-P-O}$	[1]	mdynÅ/rad^2
Interaction terms				
$K_{O-Si,Si-O'}$[2]	0.711	$K_{O-Al,Al-O'}$[2]	0.594	mdyn/Å
		$K_{O-P,P-O'}$[2]	0.195	mdyn/Å
$K_{Si-O,O-Si'}$	0.843	$K_{Al-O,O-P}$	−0.438	mdyn/Å
$K_{O-Si,O'-Si-O''}$[2]	−0.263	$K_{O-Al,O'-Al-O''}$[2]	−0.014	mdyn/rad
		$K_{O-P,O'-P-O''}$[2]	−0.122	mdyn/rad
$K_{Si-O,Si-O-Si'}$	0.398	$K_{Al-O,Al-O-P}$	[1]	mdyn/rad
		$K_{P-O,Al-O-P}$	[1]	mdyn/rad
$K_{O-Si-O,O'-Si-O''}$[2]	−0.167	$K_{O-Al-O,O'-Al-O''}$[2]	−0.099	mdynÅ/rad^2
		$K_{O-P-O,O'-P-O''}$[2]	−0.272	mdynÅ/rad^2

[1] This term is not used in the AlPO$_4$ GVFF potential.

[2] This interaction is calculated for internal coordinates with only the T–atom in common.

whether particular structural groups or ring structures in zeolites cause character-istic infrared or Raman active crystal modes[96−98]. An early assignment of vibra-tional modes in zeolites is given in Table 4. The first distinction is between spectral bands that can be used to determine the type of zeolite (inter–tetrahedral modes) and bands that are almost the same for all zeolites (intra–tetrahedral modes). This classification has been used extensively in zeolite structure analysis[98,102] and crystallization studies[116−118]. Some authors[119−121] tried to distinguish AlO$_4$

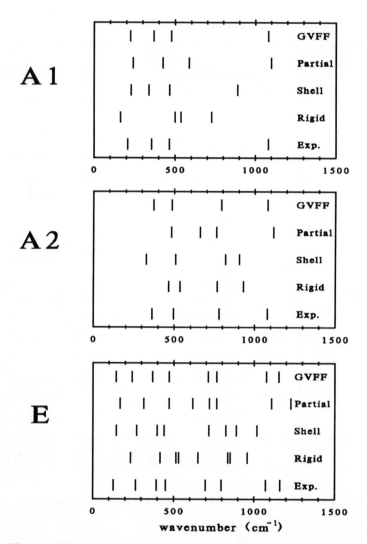

Figure 2 Spectra of α–quartz, calculated using Etchepare's GVFF, the Partial Charge model, the Shell model and the Rigid Ion model, sorted on irreducible representations for $\vec{k} = 0$. Experimental data from Refs. 114 and 115.

from SiO_4 intra–tetrahedral modes in the spectrum but this cannot be sustained expfferimentally[122].

Table 4.   Assignment of infrared zeolite bands according to Flanigen[94].

Intra–tetrahedral (structure insensitive)	
Asymmetrical stretch	$950 — 1250$ cm^{-1}
Symmetrical stretch	$650 — 720$ cm^{-1}
O–T–O bend	$420 — 500$ cm^{-1}
Inter–tetrahedral linkages (structure sensitive)	
Double ring	$500 — 650$ cm^{-1}
Pore opening	$300 — 420$ cm^{-1}
Symmetrical stretch	$750 — 820$ cm^{-1}
Asymmetrical stretch	$1050 — 1150$ cm^{-1}

A theoretical analysis of vibrational spectra of extended systems indicates that coupling causes a loss of the characteristics of vibrational modes of isolated clusters when they are connected in a Bethe lattice[67,100]. This questions structure–spectra relations such as those expressed in Table 4.

Early calculations on vibrational spectra of zeolitic systems were performed on molecular substructures[101] that were simply isolated from the crystal. Some authors slightly adapted the molecular structures to simulate better the surroundings of a smaller substructure in a real zeolite, (e.g. the pseudo lattice method[64]). However, for a clear understanding of structure–spectra relations one has to perform spectra calculations on periodic systems, not on molecules. The calculations presented here are done with the VIBRAT program[123] which uses the harmonic approximation and applies to zero wave vector modes. Eigenvectors, i.e. atomic displacement vectors, as well as eigenvalues, i.e. frequencies, of the dynamical matrix are calculated. The crystal symmetry is used explicitly to reduce the dynamical matrix. The silica GVFF force field is used, so we are not hindered by differences in aluminum contents or locations between the various zeolites.

For the calculation of the infrared line intensity I_k of the kth mode one usually takes a very simple approximation[123,124] (Eq. 5).

$$I_k = \left| \sum_i q_i \vec{u}_{ki} \right|^2 \tag{5}$$

In Eq. 5 i runs over all atoms, q_i is the (formal) charge of the ith atom and $\vec{u}_{ik}$ is the displacement vector of the ith atom in the kth mode. In a more elaborate scheme the effects of bonds are included by replacing the scalar charge q_i by an effective charge tensor $\mathbf{B}_i$ which is defined in Eq. 6[114].

$$\mathbf{B}_i = q_i \mathbf{I} - Q_i \sum_j{}^i \frac{\vec{r}_{ij}\, \vec{r}_{ij}}{|\vec{r}_{ij}|\, |\vec{r}_{ij}|} \tag{6}$$

q_i is the ionic charge of atom i, Q_i is its covalent charge, $\mathbf{I}$ is the unit tensor, $\vec{r}_{ij}$ is the bond vector between atoms i to j, $\vec{r}_{ij}\, \vec{r}_{ij}$ is a diad and $\sum^i$ is the sum over the nearest neighbors of atom i. In the case of a shell model one should, for each vibrational mode, displace the cores according to the eigenvector of the mode, relax the shell positions in the deformed structure (the shells are massless and can therefore accommodate immediately to the change in structure during the lattice vibration) and then calculate the dipole change caused by the displacements of the core and shells. Equations 5 and 6 apply to unpolarized infrared spectra.

The derivation of a simple formula for Raman line intensities R_k has attained a lot of discussion[114,125−128]. The approximation used here is taken from Ref. 123. For a specific vibrational mode k the Raman line intensity can be approximated by Eq. 7.

$$R_k = \frac{A\alpha_k^2 + B\beta_k^2}{45} \tag{7}$$

A and B are empirical constants (values of 45 and 0.1 are taken, based on empirical factors used for silicate glasses), α and β are the spherical part of the polarizability and the anisotropy, respectively, as given in Eqs. 8 and 9.

$$\alpha_k = \frac{1}{3}\Big((\rho_k)_{11} + (\rho_k)_{22} + (\rho_k)_{33}\Big) \tag{8}$$

$$\beta_k = \frac{1}{2}\Big(\big((\rho_k)_{11} - (\rho_k)_{22}\big)^2 + \big((\rho_k)_{11} - (\rho_k)_{33}\big)^2 + \big((\rho_k)_{22} - (\rho_k)_{33}\big)^2\Big) \tag{9}$$

The tensor ρ_k is the so called differential bond–polarizability tensor, defined as a diad in Eq. 10.

$$\rho_k = \sum_j F_j\, (\overrightarrow{\Delta r}_{k,j})\, (\overrightarrow{\Delta r}_{k,j}) \tag{10}$$

The summation in Eq. 10 runs over all bonds j. F_j is a bond–polarizability factor which only depends on the types of atoms forming the bond. The vector $\overrightarrow{\Delta r}_{k,j}$ is given by Eq. 11

$$\overrightarrow{\Delta r}_{k,j} = \vec{r}_j \frac{|\,\vec{r}_j + \vec{u}_{k,j1} - \vec{u}_{k,j2}\,|}{|\,\vec{r}_j\,|} \tag{11}$$

using the equilibrium bond vector $\vec{r}_j$ and the displacement vectors $\vec{u}_{k,j1}$ and $\vec{u}_{k,j2}$ of the two atoms forming the jth bond. In principle Raman spectra can be calculated for any experimental geometry (directions and polarization of in– and outgoing beam relative to the crystal axes) but since most zeolite spectra are obtained from powder samples merely averaged spectra are useful.

Because the calculations are made in the harmonic approximation no information about line widths is obtained. To facilitate the comparison between calculated and experimental spectra a Gaussian line shape[104] with 10 cm^{-1} full width at half of the maximum is assumed for each peak, both in the infrared and the Raman spectra. Significantly higher line widths, up to 75 cm^{-1}, have been observed[104] but because of our interest in specific modes a moderate line width is taken.

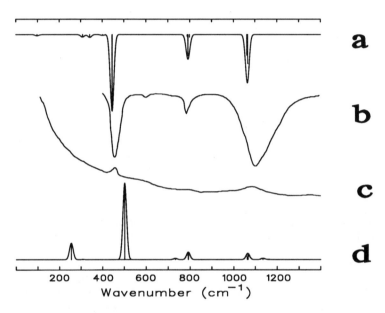

Figure 3 Spectra of silica–sodalite.
  a: Calculated infrared spectrum;   b: Experimental infrared spectrum;
  c: Experimental Raman spectrum;   d: Calculated Raman spectrum.

Figure 3 shows calculated and measured silica–sodalite infrared and Raman spectra and Figure 4 shows the same for dealuminated faujasite. The calculated and experimental spectra agree reasonably so the force field, which was developed for dense silica structures like α–quartz, can be transferred to more open structures such as zeolites. This is an indication of the highly covalent character of these silica systems.

For the analysis of the presence of substructure sensitive modes three methods have been applied[129]:

1: Simulated spectra of zeolites and molecular substructures are compared directly.

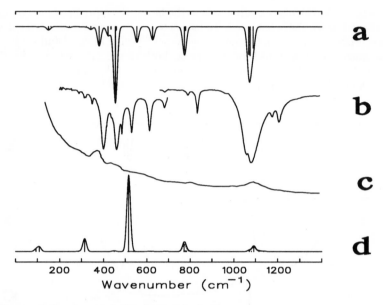

Figure 4 Spectra of dealuminated faujasite.
  a: Calculated infrared spectrum;   b: Experimental infrared spectrum;
  c: Experimental Raman spectrum;   d: Calculated Raman spectrum.

2: The summation spectrum of some zeolites containing a specific substructure is compared with the summation spectrum of zeolites without that substructure.
3: The vibrational modes of a molecular substructure are compared with those of the zeolite crystal by using the eigenvectors of the vibrational, rotational and translational modes of the substructure molecule as a complete set of basis vectors on which the eigenvectors of the vibrational modes of one of the corresponding substructures in the crystal are projected out after putting both structures in the same orientation. The projection is carried out for every vibrational mode of the crystal. The squared length of the crystal eigenvector projected on the eigenvector of a certain molecule mode is a measure of the correspondence between the crystal mode and the specific molecule mode.

Spectra of substructures, regarded as free molecules, are given in Figure 5, together with the calculated crystal spectra shown previously. The structure formulas are Si_nO_{3n} for the single n–rings and $Si_{2n}O_{5n}$ for the double n–rings* so the rings consist of complete SiO_4 tetrahedra. Because the GVFF does not include Coulomb terms there are no charge neutrality constraints on the molecular structures.

---

*   Only T–atoms are counted, see Figure 18.

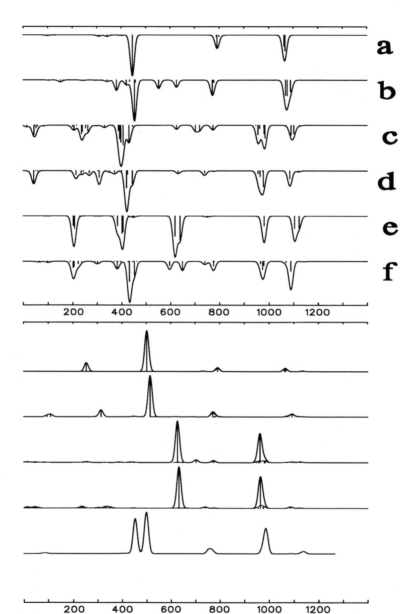

Figure 5 Calculated infrared (top) and Raman (bottom) spectra.
a: sodalite;   b: faujasite;   c: single four ring;
d: single six ring;   e: double four ring; f: double six ring.

The faujasite structure (Figure 1) contains single four rings, single six rings and double six rings while the sodalite structure contains only single four rings and single six rings. The geometries of the molecular rings are assumed to be the same as in the faujasite crystal. The double four ring is extracted from the D.L.S. structure of siliceous zeolite $A^{(130)}$. Whereas in some cases some features of the ring structure spectra appear to correspond to zeolite spectral features there is no clear correspondence between substructure spectra and spectra of crystals containing the same substructure. This shows that one cannot neglect the coupling between the vibrations of the substructures in the crystal. However, this still does not exclude the possibility of crystal vibrations that indicate the presence of such substructures but are present at frequencies not found in molecular spectra. In Figure 5 only the infrared and Raman active modes are shown. The selection of group theoretically active modes for the molecules when comparing molecules and crystals can be discussed: non-active substructure modes might have a large contribution to active lattice modes. So the lack of correspondence between crystal and substructure spectra does not directly imply that spectra of crystals containing the same substructures cannot be similar. Therefore calculated spectra of complete crystals have been compared as well, analogous to the experimental work of Flanigen[94,95]. For our analysis we will focus on double six rings.

In Figure 6 calculated infrared spectra of zeolites containing a double six ring are given. Notwithstanding some cases of correspondence between the spectra, they cannot be considered to show unique features due to the presence of double six rings. The peaks at about 1100 cm^{-1} and 800 cm^{-1} are found in all zeolites. Their exact location is determined by the Si$-$O$-$Si angle[67,100]. The band around 450 cm^{-1} in Figure 6 is mainly of O$-$Si$-$O bending character and is also found in any zeolite.

The double ring mode band suggested by Flanigen et al. (Table 4) is close to the large band to be assigned to bending modes. However, in the ring mode region of the calculated spectra no single peak common to all structures is found. The large similarity between the spectra of faujasite and Breck's Structure Six (BSS, the hexagonal form of faujasite) that has been observed experimentally[135] is reproduced by our calculations.

A summation of spectra of some double six ring containing zeolites is shown in Figure 7, a summation of spectra of some zeolites without double six rings is presented in Figure 8. The summation of experimental spectra obtained by Flanigen[94] is also given in these figures. It is obvious that no double six ring specific peak or band is found in the calculated or the experimental spectra. It should be noted that the experimental spectra are taken from aluminum containing zeolites. This partially explains the huge width of the high frequency band and the lack of sharp peaks in the summation spectra compared with the calculations.

The result of the projection of the crystal double six ring modes of faujasite on the modes of the corresponding free molecule is given in Figure 9. Only group theoretically infrared active crystal modes (T_{1u}) are drawn. For faujasite some high projection values are found. The corresponding modes in the high frequency region are not useful for structure determination because all zeolites show infrared activity

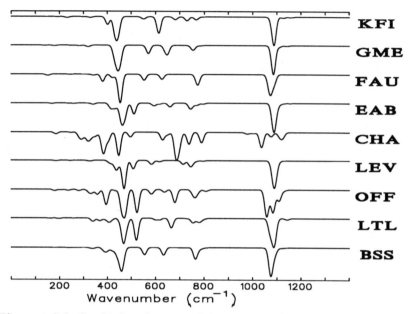

Figure 6 Calculated infrared spectra of double six ring containing zeolites: ZK-5[131], gmelinite[130], faujasite[130], E(AB) [130], chabazite[132], levyne[130], offretite[133], Linde type L[130], and Breck's Structure Six[134] (References point to structural data used).

in this part of the spectrum. In the spectral region between 500 and 650 cm^{-1} the crystal mode at 627 cm^{-1} is found having considerable double–six ring character, in agreement with earlier suggestions by Flanigen[94], although this is not confirmed by the infrared spectra calculations.

The projection of the infrared active crystal vibrations on the vibrations of a free 12–ring with the same structure as the faujasite pore is given in Figure 9. Again some large projection values are found. In the bending region the crystal modes at 453, 421 and 245 cm^{-1} correlate to some extent with molecular 12–ring modes. However, the atomic displacement vectors of these modes do not represent a symmetric pore opening mode, as suggested by Flanigen[94], and the highest projections are found for modes at 1075 and 697 cm^{-1} so even this large substructure is hardly detectable by infrared spectroscopy.

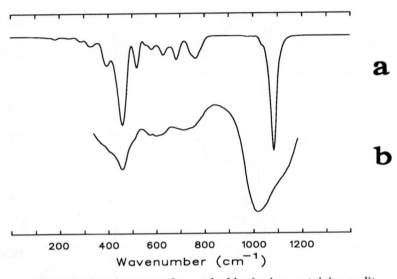

Figure 7 Sum of infrared spectra of some double six ring containing zeolites. **a**: Calculated; **b**: Experimental.

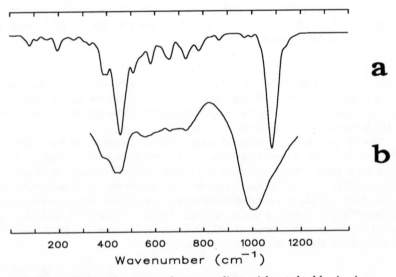

Figure 8 Sum of infrared spectra of some zeolites without double six rings. **a**: Calculated; **b**: Experimental.

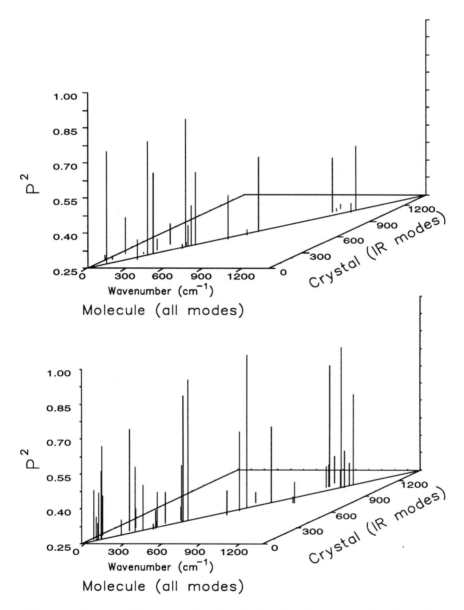

Figure 9 Square of the projection of faujasite vibrations on modes of a molecular double six ring (top) and 12–ring (bottom). Only projection values for infrared active crystal modes are given.

5. Structure refinement.

An example of the combination of experimental data and atomistic modelling in structure determination is the refinement of the structure of the clathrasil dodecasil–3C[136]. A picture of dodecasil–3C according to its first experimental X.R.D. structure refinement[6] is given in Figure 10. Next to X.R.D. data also elastic constants[137] and ^{29}Si N.M.R.[138-141] and infrared[136] spectra have been measured.

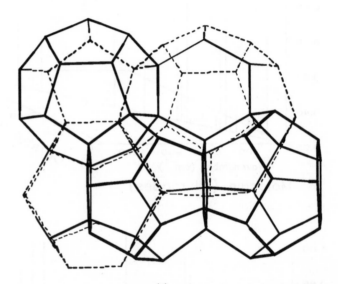

Figure 10 Structure of dodecasil–3C,[6] showing four dodecahedral cages and a hexadecahedral cage (projected alomost parallel to the dodecahedra layer).

Some results of relaxing the structure with the Rigid Ion, Shell and Partial Charge model are given in Table 5. In this table A expresses the anisotropy of the elastic constants (Eq. 12). An A value of one indicates isotropy.

$$A = C_{max}/C_{min} \qquad (12)$$

with C_{max} = maximum C'_{11} value for any direction
      C_{min} = minimum C'_{11} value for any direction

The symmetry (Fd$\bar{3}$) of the experimental structure is preserved by the relaxation but a large discrepancy between the measured and calculated elastic constants can be observed. Calculations of vibrational modes of the relaxed structures using

Table 5. Structure, lattice energy and elastic constants of cubic dodecasil–3C.

	Experiment	Rigid Ion	Shell	Partial Charge
	Lattice energy (kJ/mol SiO$_2$)			
	—	−11975.3	−12402.9	−5591.60
	Average Si−O distance (Å)			
	1.566 [6]	1.604	1.585	1.606
	Average Si−O−Si angle (degree)			
	174.5 [6]	176.5	174.3	176.1
	Unit cell (Å)			
	19.402 [6]	19.893	19.627	19.918
	Elastic constants (10^{11} dyn/cm^2)			
C$_{11}$	5.5 [137]	15.994	19.216	20.114
C$_{12}$	1.1 [137]	5.592	10.755	11.812
C$_{44}$	2.4 [137]	5.222	4.195	4.052
K$_0$	2.6 [137]	9.06	13.58	14.58
A	1.1 [137]	1.002	1.003	1.007

the corresponding potential sets showed some imaginary frequencies, in other words, the structures were not completely relaxed but remained on a saddle point in energy hyperspace. New relaxations were performed that started with structures that were obtained by deforming the saddle point structures according to atomic displacements of the vibrational modes with imaginary frequencies. The structures obtained this way (the double relaxed structures) have elastic constants that are in better agreement with the experimental values (Table 6). The double Rigid Ion model relaxation again ended in the original symmetry, and is therefore excluded from the table, but the Shell and Partial Charge model relaxation lowered the symmetries significantly: the Shell model ended with a triclinic and the Partial Charge model with a monoclinic structure. The phonon spectra of the double relaxed structures did not contain imaginary frequencies, indicating that stable minima were reached. It has been suggested[39] not to restrict the double relaxation to structures deformed by modes with imaginary frequencies but to include modes with a low frequency (< 50 cm^{-1}) as well.

A closer look at the experimental atomic coordinates shows that the cubic structure contains Si−O−Si angles which are forced to be straight by the cubic symmetry (the corresponding oxygen atoms (designated O4) lie on a threefold axis).

Table 6.   Lattice energy, structure, and elastic constants after double relaxation of cubic dodecasil–3C.

	Experiment	Shell[1]	Partial Charge
	Lattice energy (kJ/mol SiO_2)		
	—	−12408.1	−5592.0
	Symmetry		
	$Fd\bar{3}$	$P\bar{1}$	C2/c
	Average Si−O distance (Å)		
	1.565	1.598	1.606
	Average Si−O−Si Angle (degrees)		
	174.5	151.5	169.6
	Unit cell (Å)		
c	19.402	19.051	19.752
a/c	1.0000	1.0070	1.0000
b/c	1.0000	1.0044	1.0065
	Elastic constants[2] (10^{11} dyn/cm^2)		
C_{ii}	5.5	7.919	9.279
C_{ij}	1.1	1.269	1.385
C_{jj}	2.4	3.317	3.491
K_0	2.6	3.48	3.61
A	1.1	1.242	1.597

[1] The Shell model structure is triclinic with α=89.47°, β=89.87° and γ=90.07°.

[2] C_{ii} is the average of C_{11}, C_{22} and C_{33}; C_{jj} is the average of C_{44}, C_{55} and C_{66}; and C_{ij} is the average of C_{12}, C_{13} and C_{23}. The Shell model structure has even more non–zero off–diagonal elastic constants.

Another remarkable feature of the experiment is revealed by the ellipsoids that can be derived from the experimental anisotropic temperature factors (Figure 11). The axes of the ellipsoid of the O4 atoms are much larger than of any of the other ellipsoids, except for the axis in the Si−O4−Si direction. This might be an effect of dynamic disorder (the O4 atom is less apt to move in the Si−Si direction) but is more probable due to static disorder. This means that the threefold axis has to be omitted as was already indicated by the double relaxations. The conclusion can be confirmed by a simulation of the thermal ellipsoids which describe the average thermal motion of the atoms and are part of the anisotropic temperature factors.

The calculation of the thermal ellipsoids is done by a diadic summation of the atomic displacement vectors for a representative set of phonon wave vectors. For atom i the mean square displacement matrix U_i is given by Eq. 13.

$$(U_i)_{\alpha\beta} = \sum_{\vec{k}} \sum_{j=1}^{3n} \frac{\hbar w_{\vec{k}}}{2M_i \omega_{(\vec{k};j)}} \; (\vec{v}_{(\vec{k};j;i)})_\alpha \; (\vec{v}_{(\vec{k};j;i)})_\beta \; \coth\left(\frac{\hbar\omega}{2k_B T}\right) \qquad (13)$$

with $\quad \vec{v}_{(\vec{k};j;i)}$ = displacement vector of atom i for mode $(\vec{k};j)$:

$$\sum_{i=1}^{n} |\vec{v}_{(\vec{k};j;i)}|^2 = 1$$

α, β = x, y, z
n = number of atoms
$\vec{k}$ = wave vector
$w_{\vec{k}}$ = weight of wave vector (depending on integration method); $\sum_{\vec{k}} w_{\vec{k}} = 1$
$\omega_{(\vec{k};j)}$ = frequency of mode $(\vec{k};j)$
M_i = mass of atom i
k_B = Boltzmann constant
T = temperature (273 K)

The summation runs in each dimension over N sample points using an "uneven" Brillouin zone sampling method[142] to prevent divergence of U due to acoustic modes near the Brillouin zone center. The components of the ith wave vector $\vec{k}_i$ are given in Eq. 14.

$$(\vec{k}_i)_\alpha = \frac{1/2 \left(\frac{i^2}{3+i^2}\right)^{1/3} + \sum_{j=1}^{i-1} \left(\frac{j^2}{3+j^2}\right)^{1/3}}{S} \qquad (i = 1...N) \qquad (14)$$

with

$$S = \sum_{j=1}^{N} \left(\frac{j^2}{3+j^2}\right)^{1/3}$$

A grid of $9 \times 9 \times 9$ points in the positive octant of the Brillouin zone was applied. No use of symmetry was made because double relaxed structures were used. In Figure 12 the calculated thermal ellipsoids of the O4 atom in the double relaxed dodecasil-3C structures are compared with the experimental ellipsoid. The Partial Charge model gives an almost globular ellipsoid that is much smaller than the Shell model one. This is probably due to the fact that the lower vibrational frequencies calculated by the Partial Charge model are too high, as has already been seen in the vibrational spectra of α–quartz (Figure 2). Frequencies that are too high result

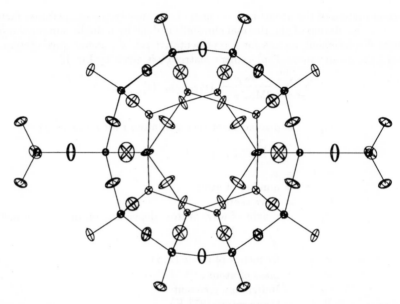

Figure 11 Ellipsoids given by experimental anisotropic temperature factors.[6]

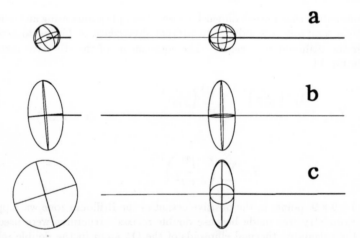

Figure 12 Thermal ellipsoid of the oxygen atom O4 in two projections: along the Si−Si line (left) and perpendicular to it (right).
a: Calculated by the Partial Charge model (magnified by a factor of two);
b: Calculated by the Shell model (magnified by a factor of two);
c: From experimental anisotropic temperature factors.

in a systematical underestimation of the amplitudes of the vibrational displacement vectors (Eq. 13). The absence of three–body interactions (bond bending terms) in the Partial Charge model might even more lower the quality of the low frequency vibrational modes that make up the main contributions to the ellipsoids.

More recent experimental structure refinements of dodecasil–3C indicated a tetragonal symmetry[8,139,140] and orthorhombic, monoclinic and triclinic symmetries[143] at low temperatures. In these structures the 180° Si–O–Si angle has disappeared, although some wide angles remain.

For a simulation of ^{29}Si N.M.R. spectra an empirical relation[144] between the chemical shift δ (in ppm relative to TMS) and the mean Si–O–Si angle per silicon atom can be used (Eq. 15):

$$\delta = -48.61 \sec (\overline{Si - O - Si}) - 168.04 \qquad (15)$$

Spectral line widths of 0.5 ppm (FWHM) were used for all lines. The simulated N.M.R. spectrum of the doubly relaxed Shell model structure shows a good agreement in the number of peaks with the experimental[141] spectrum of dodecasil–3C at 213 K (Figure 13) while the doubly relaxed Partial Charge model structure corresponds to the high temperature (373 K) dodecasil–3C. The exact location of the peaks, however, is not simulated very well. This is partially due to the very simple method used for calculating N.M.R. shifts but there is also a large difference between Si–O–Si angles in the two calculated structures: the Shell model structure has a mean Si–O–Si angle of 152° while the Partial Charge model structure shows a value of 170°. The experimental low temperature spectrum shows a peak below the 180° limit of the simulation formula, indicating that a simple dependence of N.M.R. shifts on Si–O–Si angles cannot describe all features of N.M.R. spectra.

The combination of X.R.D., N.M.R. spectroscopy, the measurement of elastic constants and modelling has lead to a reassessment of the original dodecasil–3C structure and has shown that its symmetry has to be lowered. Experiments[8,143] suggest the existence of a large number of dodecasil–3C phases. Free energy lattice minimizations can become an important tool for further research on this interesting clathrasil structure.

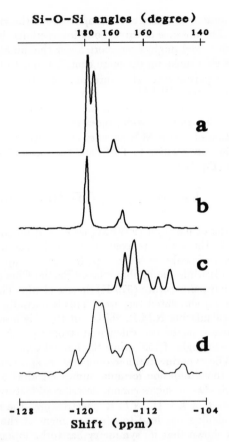

Figure 13 N.M.R. spectra of dodecasil–3C.
a: Calculated from the doubly relaxed Partial Charge model structure;
b: Experiment[141] at 373 K;
c: Calculated from the doubly relaxed Shell model structure;
d: Experiment[141] at 213 K.

6. Hypothetical structures.

Next to the refinement of existing structures atomistic modelling can also be used to predict the relative stability of hypothetical structures. One aim of current zeolite synthesis research is to increase the size of the micropores. Hypothetical structures of zeolitic compounds with wide pores can be designed by procedures as proposed by Smith[145] and Meier[146]. An approach has been suggested that requires substitution of small structural elements by larger elements showing the same external connectivity. Sometimes these large substitutional elements contain rings consisting of three tetrahedra, as is illustrated in Figure 14 in which one tetrahedron is substituted by a tetrahedron of SiO_2 tetrahedra, the so called 3^4 unit[145]. The result of applying this method to sodalite[147] is shown in Figure 15. We will call this structure 3^4–sodalite. The 3^4 unit is about twice as large as the ideal tetrahedron, resulting in a decrease in the framework density (expressed in T–atoms per 1000 Å^3) to $4 \times 2^{-3} = 0.5$ times the original value upon substituting all tetrahedra by 3^4 units. A correlation between the low density of a structure and the presence of low membered rings has been shown by Brunner and Meier[148]. Recently an enumeration of three–ring containing hypothetical structures has been given by Bosmans and Andries[149].

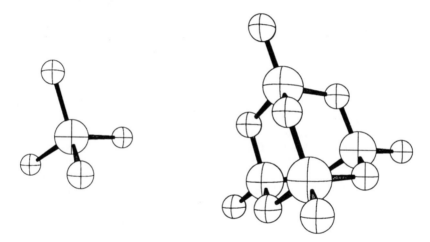

Figure 14 The elementary tetrahedron compared with the 3^4 unit.

Ab–initio calculations[13,150] with a STO–3G basis set of fully relaxed ring structures of $[SiO(OH)_2]_n$ indicated strain in the three–ring of the order of 9 kJ/mol $SiO(OH)_2$ and an absence of strain in the four and five rings (all relative to a six–ring). Calculations with a 3–21G basis set[79] give an energy difference of 29.2 kJ/mol SiO_2 between a three–ring and a five–ring. The strain in the three–rings also inhibited the formation of three–rings in computer simulations of vitreous silica[151]. So one can question the possible existence of siliceous zeolitic systems with three–rings.

There exist three zeolitic species containing three–rings: lovdarite[152,153], a structure with silicon and beryllium framework atoms; VPI–7, a zincosilicate[154]

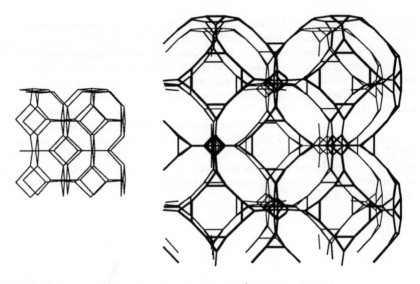

Figure 15 Structures of sodalite (left) and 3^4–sodalite (right).

with a structure resembling lovdarite, and the alumino–silicate ZSM–18[155]. For lovdarite two structures (lovdarite I and lovdarite II) have been proposed[153]. For ZSM–18 three models (A, B and C) are suggested which will be compared with a silica analogue of ALPO–46[156] because their structures show some resemblance.

Lattice energy minimizations are performed for pure silica analogues of these structures[157]. The results are shown in Table 7 together with those of some other zeolites. The lattice energies calculated for different structures can only be compared when the same potential model is used. The partial charge model shows a spurious dependence of the lattice energy on framework density which should be corrected[79] by the formula in Eq. 16.

$$E_c = E + 4.55639 \cdot n_{fr} \qquad \text{for } SiO_2$$
$$E_c = E + 3.79207 \cdot n_{fr} \qquad \text{for } AlPO_4 \qquad (16)$$

with E_c = corrected energy (kJ/mol TO_2);

E = calculated energy (kJ/mol TO_2);

n_{fr} = framework density (T − atoms/1000 $\overset{\circ}{A}^3$).

The lattice energies of the ZSM–18 structures are close to that of faujasite, as is the framework density. ZSM–18 model C seems to be most favorable since

Table 7.   Experimental densities, lattice energies of relaxed structures (relative to α–quartz) and three–ring contribution.

Structure	Density[2] (T/1000Å³)	Lattice energy (kJ/mol SiO₂)				n_3[1] %
		Rigid Ion	Shell	Partial Charge[3]		
faujasite	13.5	41.7	19.5	64.7	4.5	—
VPI–7	17.2	53.6	23.5	63.0	34.8	56
lovdarite I	18.3	54.6	24.0	66.8	25.3	56
lovdarite II	18.3	54.1	23.6	63.1	30.3	56
ZSM–18 A	14.7	43.6	18.7	61.8	7.5	18
ZSM–18 B	14.3	36.6	17.7	61.1	5.9	18
ZSM–18 C	15.3	33.4	15.7	54.6	4.0	33
Si–ALPO–46	13.7	38.9	17.9	61.2	6.0	—
3^4–sodalite	8.6	214.4	102.6	157.7	72.5	100
sodalite	17.1	0.6	13.0	47.0	3.7	—

[1] n_3 is the relative number of T–atoms being part of a three–ring.

[2] Density based on experimental, non–siliceous, structure.

[3] For the Partial Charge model E and E_c are given (see formula 16).

this model has the lowest lattice energy. Experimental X.R.D. data indicate that model B is more likely[155]. Model C has the highest framework density of the three ZSM–18 models. Our calculations apply to pure silica systems while the experiments concern aluminum containing samples. In general low density zeolites have to be stabilized by framework aluminum[31,158] while high density zeolites can exist in a high silica form, so our calculations tend to lead to high density phases. The computed lattice energy and framework density of silica–ALPO–46 are very close to those of ZSM–18. So in this case the introduction of three–rings does not lead to a significant decrease of framework density or framework stability. The lattice energy of lovdarite is higher than the energy of ZSM–18 because lovdarite has relatively more T–atoms participating in a three–ring than ZSM–18 (see the last column of Table 7). This trend continues for 3^4–sodalite, which is energetically unfavorable, having a very low lattice energy which leaps out of the range of values for silica polymorphs found before[69,158]. The instability of 3^4–sodalite is caused by the distortion of the SiO₄ tetrahedra in the 3^4 unit. Quantum–chemical calculations of Kramer[79] also indicate the improbability of the 3^4 unit. The calculated energy difference between the 3^4 unit and the double six ring ($4^6 6^2$ unit) is 99.6 kJ/mol Si using a 3–21G basis set. Table 8 lists interatomic angles and distances in three–rings obtained from two ab–initio calculations and average values for all our energy minimized structures. It is clear that the angles of the (molecular) ab–initio geometry of a three–ring agree well with values calculated with the Shell model. Although the Rigid Ion model has a O−Si−O three–body interaction term, it gives a larger deformation of the tetrahedral angle than the Partial Charge model, which has no explicit O−Si−O term.

Table 8. Angles and distances in silica three–rings.

	Ab–initio		Rigid Ion	Shell	Partial Charge
Si−O (Å)	1.644[1]	1.622[2]	1.618	1.630	1.621
O−Si−O	109.4	105.7	96.38	108.03	102.01
Si−O−Si [3]	124.9	130.2	142.70	129.51	137.32

[1] STO–3G geometry optimized $(H_2SiO)_3$ with C_{3v} symmetry[150].
[2] STO–3G geometry optimized $(H_2SiO_3)_3$ with C_{3v} symmetry[13].
[3] The experimental value for cyclo–silicates[150] is 126°.

Next to structural information the infrared and Raman spectra of three–ring containing compounds are of interest. In section 4 we showed that specific substructures are difficult to detect in these spectra. The special nature of the three–rings might cause them to be an exception to this observation. Figure 16 shows a selection of calculated spectra for structures with and without three–rings[157]. The structures were relaxed with the Shell model and the spectra were calculated with the silica GVFF.

Studies on vitreous silica indicate the presence of a three–ring band at 606 cm^{-1} (the so called D_2 band[151]). The modes around 750 cm^{-1} can be explained completely by the Bethe lattice method from the small Si−O−Si angles in three–ring containing structures. Differences between calculated infrared spectra of structures with and without three–rings are not very prominent so three–rings seems to conform to the conclusions of section 4. The Raman spectra of sodalite and 3^4–sodalite, however, show some large differences, which are not consistent with the other spectra.

Another case in which some topologically distinct structures have been proposed with almost the same unit cell dimensions is the structure determination of the AlPO$_4$ VPI–5[159]. This AlPO$_4$ has a very wide pore (a ring with 18 T–atoms and a diameter of 12 Å) and is member of a set of closely related 18–ring structures enumerated by Richardson et al.[160]. The lattice energies of all members of the set are given in Table 9. It indicates that hexagonal prisms and cubes in the structure increase the lattice energy, thus decreasing the plausibility of the structure. This is not the only energy determining feature, the relative orientation of neighboring

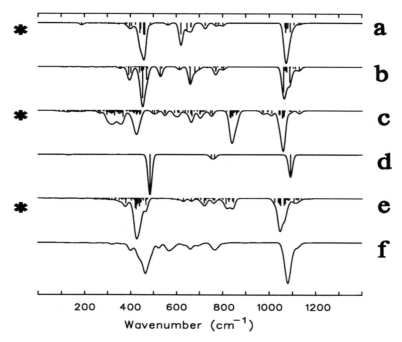

Figure 16 Calculated infrared spectra of silica zeolites relaxed with the Shell model. Spectra of structures with three–rings are marked with an asterisk.

   a: ZSM–18 model B;   **b**: silica analogue of ALPO–46;

   c: hypothetical 3^4–sodalite;   **d**: sodalite;

   e: lovdarite I;   **f**: average of zeolites without three–rings.

tetrahedra has also an effect[161]. The presence of a crankshaft chain (Figure 18) seems to make a structure more favorable.

The VPI–5 topology is part of an infinite range of hexagonal (4;2) networks with increasing pore size originally designated 81, 81(1) and 81(2)[145,162] *etcetera* (Figure 19) with pore sizes of 12, 18 and 24 T–atoms, respectively. Another 24–ring structure, here referred to as Super–VPI–5, can be made from the 18–ring structure by replacing the crankshaft common to the two four–rings in VPI–5 by a double crankshaft. A double crankshaft has the same external connectivity as a single one, so this procedure could be repeated infinitely. The calculated differences in lattice energy between the 12–ring structure AlPO$_4$–5[110], the 18–ring structure VPI–5 (net 520) and the hypothetical 24–ring structure Super–VPI–5 are mainly due to density differences, no direct effect of the pore size on the lattice energy has been observed.

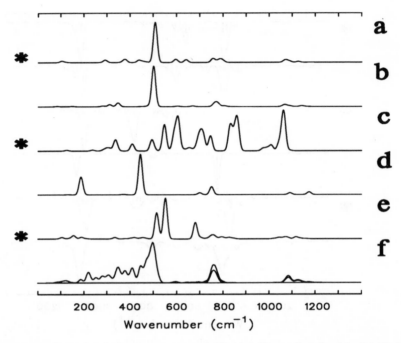

Figure 17 Calculated Raman spectra of silica zeolites relaxed with the Shell model. Spectra of structures with three–rings are marked with an asterisk.

a: ZSM–18 model B;   b: silica analogue of ALPO–46;

c: hypothetical 3^4–sodalite;   d: sodalite;

e: lovdarite I;   f: average of zeolites without three–rings;

The unit cells of $AlPO_4$–5, VPI–5 and Super–VPI–5 comprises two layers. In the experimental structure refinements it is assumed that they are stacked exactly on top of each other. The lattice relaxations, however, show a shift of layers for $AlPO_4$–5 and VPI–5 while Super–VPI–5 keeps the layers stacked. This can be explained by a higher flexibility of the 24–ring. Whereas the strain in the initial Super–VPI–5 structure can be accommodated by a deformation of the 24–rings (resulting in a very low symmetry) the rings in $AlPO_4$–8 and VPI–5 are not changed and the reduction of energy has to be performed by the shift of layers. This effect depends strongly on the interatomic potentials[39,163] and might be useful in tuning potential parameters.

Table 9. Results of the lattice energy minimization of 18–ring nets, $AlPO_4$–5 (12–ring) and Super–VPI–5 (24–ring).

Net[160]	Rings [2]	Units [3]	Symmetry [4]		a/a_0 [5]	c/c_0 [5]	Energy[1] kJ/mol TO_2	
520	18	22a	C_{6v}^3	C_{2v}^{12}	$1.017^{[6]}$	0.999	36.37	−4.66
523	8, 18	16a	C_{3i}^1	C_{3i}^1	1.025	1.050	42.63	−1.47
521	18	18a	C_{3v}^3	C_{3v}^3	1.033	1.033	43.08	−1.02
524	18	16a	C_{3v}^2	C_{3v}^2	1.034	1.029	44.81	0.82
522	10, 18	14a 1c	C_3^1	C_1^1	$1.029^{[6]}$	0.990	44.87	3.12
525	8, 18	12a 1c	C_3^1	C_3^1	1.030	0.995	45.55	0.77
313	8, 18	8a 2c	D_{3d}^1	D_{3d}^1	1.013	1.069	45.83	−0.78
560	18	14a	C_{6v}^3	C_{6v}^3	1.046	1.032	51.06	5.82
561	14, 18	8a 2b 1c	C_3^1	C_1^1	$1.034^{[6]}$	0.938	52.00	13.25
528	10, 18	10a 1c	C_3^1	C_3^1	1.046	1.025	52.73	7.84
527	10, 18	6a 2c	C_6^6	C_6^6	1.052	0.972	54.24	11.43
526	10, 18	6a 2c	C_{3v}^2	C_{3v}^2	1.057	0.875	54.57	16.69
529	8, 12, 18	4a 2b 2c	D_3^1	D_3^1	1.031	1.091	54.92	5.85
81	12		C_{6v}^2	C_{6v}^2	1.033	1.023	27.67	−5.11
[7]	24		C_{3v}^3	C_1^1	—	—	45.95	−5.05

[1] Lattice energy according to the $AlPO_4$ Partial Charge model before and after density correction relative to the energies of α–berlinite (−5716.59 kJ/mol TO_2, corrected −5623.61 kJ/mol TO_2).

[2] Rings are described here by the number of T–atoms in the ring. Each net has four–membered and six–membered rings.

[3] Structural units per unit cell, see Figure 18.

[4] Symmetry before and after lattice energy minimization.

[5] a/a_0 and c/c_0 are the ratios between the lattice vectors of the minimized and hypothetical structures.

[6] The average of a and b is taken.

[7] The hypothetical structure Super–VPI–5.

7. Concluding remarks.

We have illustrated the current use of atomistic modelling techniques for simulating zeolitic lattices. It is shown that a combination of various experimental data and atomistic modelling enables a reassessment of lattice structures. Although the quality of the simulated vibrational spectra is limited some important conclusions have been drawn on the use of infrared spectroscopy as a structural characterization tool. Improvement of the spectra simulations requires potential sets that are especially developed for this purpose, like the GVFF. A force field that is useful

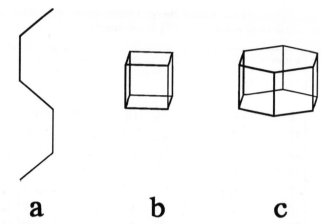

Figure 18 The three structural units given in Table 9.   **a**: crankshaft chain;
b: cube (double four ring);   **c**: hexagonal prism (double six ring).

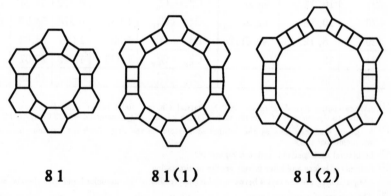

Figure 19 The nets 81, 81(1)=520 and 81(2).

for the calculation of spectra of dense systems and zeolites can be developed by
fitting high quality experimental spectra of α–quartz, β–quartz and some zeolites.
This will be facilitated by the current improvements in synthesis of large zeolite
crystallites and Raman and polarized infrared spectroscopy[164]. The method of
fitting spectra requires good structural data of the model systems used. Apart from
the application of modelling techniques also the use of synchrotron X–ray sources
has increased the quality of structural information. The need for a good description of atomic vibrations is not only found in vibrational spectroscopy but also in

the simulation of thermal ellipsoids as an aid for X.R.D. structure refinement, in free energy minimizations and in calculations on the diffusion of molecules inside zeolites.

For the simulation of chemical reactions inside zeolites a combined approach[165] of *ab–initio* quantum chemical calculations and atomistic modelling techniques has to be used to reduce the computation costs involved in quantum–chemical calculations and to avoid the parametrization problems and inaccuracies related to atomistic modelling of interactions between molecules and the zeolite lattice. In this way the changes in the lattice induced by a simple reaction, *e.g.* proton transfer, can be accounted for[165] at a level that goes beyond quantum–chemical calculations on systems in which simple point charges represent the lattice[28,29,166].

References.

1. D.W. Breck *Zeolite Molecular Sieves, Structure, Chemistry and Use* (John Wiley and Sons, Inc. New York, 1974).
2. M.E. Davis and R.F. Lobo, *Chem. of Materials* **4** (1992) 756.
3. S.T. Wilson, B.M. Lok, C.A. Messina, T.R. Cannan and E.M. Flanigen, *ACS Symp. Ser.* **218** (1983) 79.
4. H. Gies and B. Marler, *Zeolites* **12** (1992) 42.
5. H. Gies, F. Liebau and H. Gerke, *Angew. Chem.* **94** (1982) 214. H. Gies, *Nachr. Chem. Tech. Lab.* **33** (1985) 387. H. Gies, *Z. Kristallogr.* **175** (1986) 93.
6. H. Gies, *Z. Kristallogr.* **167** (1984) 73.
7. F. Liebau, H. Gies, R. Gunawardane and B. Marler, *Zeolites* **6** (1986) 373.
8. H. Chae, W. Klemperer, D. Payne, C.T.A. Suchital, D.R. Wake and S.R. Wilson *Clathrasils: new materials for non–linear optics* (ACS Symp. "New Materials for nonlinear optics", 1991).
9. E.M. Flanigen, R.L. Patton and S.T. Wilson, *Stud. Surf. Sci. Catal.* **37** (1988) 13.
10. G.H. Grant and R.J. Abrahams, *Catalysis* **8** (1990) 68.
11. G.B. Suffritti and A. Gamba, *Int. Rev. Phys. Chem.* **6(4)** (1987) 299.
12. M.D. Newton, M. O'Keeffe and G.V. Gibbs, *Phys. Chem. Minerals* **6** (1980) 305.
13. B.W.H. van Beest, J. Verbeek and R.A. van Santen, *Catal. Lett.* **1** (1988) 147.
14. J.G. Fripiat, F. Berger–André, J. André and E.G. Derouane, *Zeolites* **3** (1983) 306.
15. J. Sauer and R. Zahradník, *Int. J. Quantum Chem.* **26** (1984) 793.
16. A.G. Pelmenshchikov, E.A. Paukshtis, V.S. Stepanov, K.G. Ione, G.M. Zhidomirov and K.I. Zamaraev, *M.J. Philips (ed.): Proc. 9th Int. Conf. Cat., Canada* (1988) 404.
17. G.M. Zhidomirov and V.B. Kazansky, *Adv. in Catalysis* **34** (1986) 131.
18. G. Ooms and R.A. van Santen, *Recl. Trav. Chim. Pays Bas* **106** (1987) 69.
19. S. Beran, P. Jíru and B. Wichterlová, *J. Phys. Chem.* **85** (1981) 1951. S. Beran, *J. Phys. Chem.* **85** (1981) 1956. S. Beran, *Z. Phys. Chem.* **137**

(1983) 89. S. Beran, P. Jíru and B. Wichterlová, *React. Kinet. Catal. Lett.* **18** (1981) 51. S. Beran, *J.Phys. Chem.* **88** (1982) 111. S. Beran, *Chem. Phys. Lett.* **91** (1982) 86. S. Beran, *Z. Phys. Chem.* **130** (1982) 81.

20. G.V. Gibbs, *Am. Mineral.* **67** (1982) 421.
21. J. Sauer, *J. Phys. Chem.* **91** (1987) 2315.
22. J. Sauer and W. Shirmer, *Stud. Surf. Sci. Catal.* **37** (1988) 323.
23. V.B. Kazansky, *Bulg. Acad. of Sciences Comm. Rep. Chem.* **13** (1980) 19.
24. P.J. O'Malley and J. Dwyer, *Zeolites* **8** (1988) 317. P.J. O'Malley and J. Dwyer, *Chem. Phys. Lett.* **143** (1988) 97. P.J. O'Malley and J. Dwyer, *J. Phys. Chem.* **92** (1988) 3005.
25. W.J. Mortier, P. Geerlings, C. van Alsenoy and H.P. Figeys, *J. Phys. Chem.* **83** (1979) 855.
26. W.J. Mortier, J. Sauer, J.A. Lercher and H. Noller, *J. Phys. Chem.* **88** (1984) 905.
27. E. Kassab, K. Seiti and M. Allavena, *J. Phys. Chem.* **92** (1988) 6705.
28. R. Vetrivel, C.R.A. Catlow and E.A. Colbourn, *Stud. Surf. Sci. Catal.* **37** (1988) 309.
29. M. Allavena, K. Seiti, E. Kassab, Gy. Ferenczy and J.G. Angyán, *Chem. Phys. Lett.* **168** (1990) 461.
30. I.D. Mikheikin, I.A. Abronin, G.M. Zhidomirov and V.B. Kazansky, *Kinet. Katal.* **18** (1977) 1580. I.D. Mikheikin, I.A. Abronin, A.I. Lumpuv and G.M. Zhidomirov, *Kinet. Katal.* **19** (1978) 1050.
31. G. Ooms, R.A. van Santen, R.A. Jackson and C.R.A. Catlow, *Stud. Surf. Sci. Catal.* **37** (1987) 317. G. Ooms, R.A. van Santen, C.J.J. den Ouden, R.A. Jackson and C.R.A. Catlow, *J. Phys. Chem.* **92** (1988) 4462.
32. E. Dempsey, *J. Phys. Chem.* **75** (1969) 3660.
33. C.R.A. Catlow and W.C. Mackrodt *Computer simulation of solids* (Lecture Notes on Physics, Springer, Berlin, 1982).
34. C.R.A. Catlow, M. Doherty, G.D. Price, M.J. Sanders and S.C. Parker, *Mater. Sci. Forum* **7** (1986) 163.
35. M.J. Sanders *Computer simulation of framework structured minerals* (PhD Thesis, University of London, 1984).
36. R.A. Jackson and C.R.A Catlow, *Molec. Simulation* **1** (1988) 207.
37. A.J.M. de Man *Simulation of physical properties of zeolites* (PhD Thesis Eindhoven University of Technology, 1992).
38. K.-P. Schröder, J. Sauer, M. Leslie and C.R.A. Catlow, *Zeolites* **12** (1992) 20. K.-P. Schröder, J. Sauer, M. Leslie, C.R.A. Catlow and J.M. Thomas, *Chem. Phys. Lett.* **188** (1992) 320.
39. E. de Vos Burchart *Studies on zeolites: molecular mechanics, framework stability, and crystal growth* (PhD Thesis Delft University of Technology, 1992).
40. A.V. Kiselev, A.A. Lopatkin and A.A. Schulga, *Zeolites* **5** (1985) 261.
41. C.J.J. den Ouden *Computational studies in zeolite science; An investigation of guest-host interactions* (PhD Thesis, Eindhoven University of Technology, 1991).

42. J.O. Titiloye, S.C. Parker, F.S. Stone and C.R.A. Catlow, *J. Phys. Chem.* **95** (1991) 4038.
43. G. Aloisi, P. Barnes, C.R.A. Catlow, R.A. Jackson and A.J. Richards, *J. Chem. Phys* **93** (1990) 3573.
44. P. Demontis, G.B. Suffritti, A. Alberti, S. Quartieri and E. Fois, *Gazz. Chim. Italiana* **116** (1986) 459.
45. P. Demontis, G.B. Suffritti, S. Quartieri, E. Fois and A. Gamba, *J. Phys. Chem.* **92** (1988) 867.
46. J.G. Angyán, Gy. Ferenczy, P. Nagy and G. Naráy-Szabó, *Collect. Czech. Chem. Commun.* **53** (1988) 2308.
47. K.T. No, H. Chon, T. Lee and M.S. Jhon, *J. Phys. Chem.* **85** (1981) 2065.
48. S. Beran, *J. Mol. Catal.* **45** (1988) 225.
49. A. Goursot, F. Fajula, C. Daul and J. Weber, *J. Phys. Chem.* **92** (1988) 4456.
50. J.G. Angyán, F. Colonna–Cesari and O. Tapia, *Chem. Phys. Lett.* **166** (1990) 180.
51. W.J. Mortier and R. Vetrivel, *in: D. Barthomeuf et al.(eds.), Guidelines for mastering the properties of molecular sieves, NATO ASI 221, Plenum, New York* (1990) 263.
52. J.M. Newsam *Zeolites* (in: A.K. Cheetham and P. Day (ed.): Solid State Chemistry Vol. 2, 1991).
53. W.M. Meier and H. Villiger, *Z. Kristallogr.* **129** (1969) 411.
54. K.J. Roberts and R. Docherty, *Comput. Phys. Commun.* **64** (1991) 311.
55. S.C. Parker, C.R.A .Catlow and A.N. Cormack, *Acta Cryst.* B **40** (1984) 200.
56. M. Leslie *Daresbury Laboratory Technical Memorandum* (in preparation,).
57. G.K. Johnson, I.R. Tasker, D.A. Howell and J.V. Smith, *J. Chem. Thermodyn.* **19** (1987) 617.
58. J. Patarin, M. Soulard, H. Kessler, J.L. Guth and M. Diot, *Thermochimica Acta* **146** (1989) 21.
59. S.C. Parker and G.D. Price, *in: C.R.A. Catlow (ed.) Adv. in Solid State Chem.* **1** (1989) 296.
60. W. Löwenstein, *Amer. Mineral.* **39** (1954) 92.
61. R.G. Bell, R.A. Jackson and C.R.A. Catlow, *Zeolites* **12** (1992) 870.
62. A.J.M. de Man, B.W.H. van Beest, M. Leslie and R.A. van Santen, *J. Phys. Chem.* **94** (1990) 2524.
63. K.T. No and M.S. Jhon, *Bull. Korean Chem. Soc.* **6** (1985) 183.
64. K.T. No, D.H. Bae and M.S. Jhon, *J. Phys. Chem.* **90** (1986) 1772. K.T. No, B.H. Seo and J.M. Park, *J. Phys. Chem.* **92** (1988) 6783. K.T. No, B.H. Seo and M.S. Jhon, *Theoretica Chim. Acta* **75** (1989) 307.
65. K.T. No, J.S. Kim, Y.Y. Huh, W.K. Kim and M.S. Jhon, *J. Phys. Chem.* **91** (1987) 740.
66. J. Ariai and S.R.P.Smith, *J. Phys. C* **14** (1981) 1193.
67. R.A. van Santen and D.L. Vogel, *in: C.R.A. Catlow (ed.) Adv. Solid State Chem.* **1** (1989) 151.

68. B.W.H. van Beest, G.J. Kramer and R.A. van Santen, *Phys. Rev. Lett.* **64** (1990) 1955.
69. G.J. Kramer, N. Farragher, B.W.H. van Beest and R.A. van Santen, *Phys. Rev. B* **43** (1991) 5068.
70. S. Tsuneyuki, M. Tsukada, H. Aoki and Y. Matsui, *Phys. Rev. Lett.* **61** (1988) 869.
71. J. Purton, R. Jones and C.R.A. Catlow, Phys. Chem. Miner..
72. E. Geidel, P. Birner and H. Böhlig, *Z. Chemie* **30** (1990) 141.
73. R. Ahlrichs, M. Bär, M. Häser, C. Kölmel and J.A. Sauer, *Chem. Phys. Lett.* **164** (1989) 199.
74. A.C. Lasaga and G.V. Gibbs, *Phys. Chem. Minerals* **16** (1988) 29.
75. A.C. Hess, P.F. McMillan and M. O'Keeffe, *J. Phys. Chem.* **90** (1986) 5561.
76. G.V. Gibbs, E.P. Meagher, J.V. Smith and J.J. Pluth, *J.R. Kratzer (ed.): Mol. Sieves 2 ACS Symp. Series* **40** (1970) 19.
77. M. O'Keeffe and P.F. McMillan, *J. Phys. Chem.* **90** (1986) 541.
78. L. Stixrude and M.S.T. Bukowinski, *Phys. Chem. Minerals* **16** (1988) 199.
79. G.J. Kramer, A.J.M. de Man and R.A. van Santen, *J. Am. Chem. Soc.* **131** (1991) 6435.
80. C. Pisani, R. Dovesi and C. Roetti *Hartree–Fock ab–initio treatment of crystalline systems* (Lecture Notes in Chem., Springer, Heidelberg, 1988).
81. R. Dovesi, C. Pisani, C. Roetti, M. Causá and V.R. Saunders *CRYSTAL88, an ab–initio all–electron LCAO–Hartree–Fock program for periodic solids* (QCPE, no. 577).
82. H.H. Hummel and R.G. Gordon, *Mater. Sci. Res. Soc. Symp. Proc.* **141** (1989) 159.
83. M.J. Mehl, J.E. Osburn, D.A. Papaconstantopoulos and B.M. Klein, *Phys. Rev. B* **41** (1990) 10311.
84. R. Nada, C.R.A. Catlow, R. Dovesi and C. Pisani, *Phys. Chem. Miner.* **17** (1990) 353.
85. J.D. Gale, C.R.A. Catlow and W.C. Mackrodt, to be submitted.
86. S. Dasgupta and W.A. Goddard III, *J. Phys. Chem* **90** (1989) 7207.
87. N. Karasawa, S. Dasgupta and W.A. Goddard III, *J. Phys. Chem* **95** (1991) 2260.
88. M.J. Sanders, M. Leslie and C.R.A. Catlow, *J. Chem. Soc. Chem. Comm.* (1984) 1271.
89. B.G. Dick and A.W. Overhauser, *Phys. Rev.* **112** (1958) 91.
90. N.N. Greenwood *Ionen, Kristalle, Gitterdefekte und nichtstöchiometrische Verbindungen* (Verlag Chemie, 1973).
91. K. Iishi, M. Miura, Y. Shiro and H. Murata, *Phys. Chem. Minerals* **9** (1983) 61.
92. C.R.A. Catlow, *Proc. Royal Soc. A* **353** (1977) 533.
93. P.P. Ewald, *Ann. Phys. (Leipzig)* **64** (1921) 253.
94. E.M. Flanigen, H. Khatami and H.A. Szymanski, *Adv. Chem. Ser.* **101** (1971) 201.

95. E.M. Flanigen, *R.A. Rabo (ed.): ACS Monograph* **171** (1976) 80.
96. P.K. Dutta and B. Del Barco, *J. Chem. Soc. Chem. Comm.* (1985) 1297.
   P.K. Dutta and B. Del Barco, *J. Phys. Chem.* **89** (1985) 1861. P.K. Dutta, D.C. Shieh and M. Puri, *Zeolites* **8** (1988) 306. P.K. Dutta and B. Del Barco, *J. Phys. Chem* **92** (1988) 354.
97. L. Kubelková, V. Seidl, G. Borbély and H.K. Beyer, *J. Chem. Soc. Faraday Trans. I* **84** (1988) 1447.
98. J.C. Jansen, F.J. van der Gaag and H. van Bekkum, *Zeolites* **4** (1984) 369.
99. P.K. Dutta, K.M. Rao and J.Y. Park, *J. Phys. Chem.* **95** (1991) 6654.
100. R.G. Buckley, H.W. Deckman, J.M. Newsam, J.A. McHenry, P.D. Persans and H. Witzke, *Mater. Res. Soc. Symp. Proc.* **111** (1988) 141.
101. C.S. Blackwell, *J. Phys. Chem.* **83** (1979) 3251. C.S. Blackwell, *J. Phys. Chem.* **83** (1979) 3257.
102. P.A. Jacobs, E.G. Derouane and J. Weitkamp, *J. Chem. Soc. Chem. Comm.* (1981) 591.
103. R. Szostak and T.L. Thomas, *J. Catal.* **101** (1986) 549.
104. A. Miecznikowski and J. Hanuza, *Zeolites* **7** (1987) 249.
105. R.A. van Santen, B.W.H. van Beest and A.J.M. de Man *On lattice dynamics, stability and acidity of zeolites* (in: D. Barthomeuf *et al.* (eds.), Guidelines for mastering the properties of molecular sieves, NATO ASI 221, Plenum, New York, 1989).
106. M.M. Elcombe, *Proc. Phys. Soc. (London)* **9** (1967) 947.
107. E.W. Kammer and J.V. Atanasoff, *Phys. Rev.* **62** (1942) 395.
108. S.D. Ross *Inorganic infrared and Raman spectra* (McGraw–Hill, London, 1972).
109. J. Etchepare, M. Marian and L. Smetankine, *J. Chem. Phys.* **60** (1974) 1873.
110. A.J.M. de Man, W.P.J.H. Jacobs, J.P. Gilson and R.A. van Santen, *Zeolites* **12** (1992) 826.
111. A.N. Lazarev *Vibrational spectra and structure of silicates* (Consultants Bureau, New York, 1972).
112. W.J. Gordy, *J. Chem. Phys.* **14** (1946) 305.
113. R.M. Badger, *J. Chem. Phys.* **2** (1934) 128.
114. D.A. Kleinman and W.G. Spitzer, *Phys. Rev.* **125** (1962) 16.
115. D. Krishnamurti, *Proc. Indian Acad. Sci. A* **47** (1958) 276.
116. B.D. McNicol, G.T. Pott and K.R. Loos, *J. Phys. Chem.* **76** (1972) 3388.
117. F. Roozeboom, H. Robson and S.C. Chan, *Zeolites* **3** (1983) 321.
118. S.T. Stojković and B. Adnadjević, *Zeolites* **8** (1988) 523.
119. H. Dutz, *Ber. Deutsch. Keram. Geselsch.* **46** (1969) 75.
120. B.D. Saksena, *Trans. Faraday Soc.* **57** (1961) 242.
121. B.H. Ha, J. Guidot and D. Barthomeuf, *J. Chem. Soc. Faraday Trans.* **75** (1979) 1245.
122. V. Stubičan and R. Roy, *Am. Miner.* **46** (1961) 32. V. Stubičan and R. Roy, *Am. Ceram. Soc.* **44** (1961) 625. V. Stubičan and R. Roy, *Z. Kristallogr.* **115** (1961) 200.
123. E. Dowty, *Phys. Chem. Minerals* **14** (1987) 67.

124. E. Dowty *VIBRAT manual* (Bogota, NJ, 1988).

125. R. Escribano, G. Del Rio and J.M. Orza, *Mol. Phys.* **33** (1977) 543.

126. M. Gussoni, *in: R.J.H. Clark and R.E. Hester (eds.): Adv. infrared and Raman spectr.* **6** (1980) 61.

127. W.B. Person and G. Zerbi (ed.) *Vibrational intensities in infrared and Raman spectroscopy* (Elsevier, Amsterdam, 1982).

128. T. Furakawa, K.E. Fox and W.B. White, *J. Chem. Phys.* **75** (1981) 3226.

129. A.J.M. de Man and R.A. van Santen, *Zeolites* **12** (1992) 269.

130. K.A. van Genechten and W.J. Mortier, *Zeolites* **8** (1988) 273.

131. W.M. Meier and G.T. Kokotailo, *Z. Kristallogr.* **121** (1965) 211.

132. J.V. Smith, F. Rinaldi and L.S. Dent Glasser, *Acta Cryst.* **16** (1963) 45.

133. J.A. Gard and J.M. Tait, *Acta Cryst. B* **28** (1972) 825.

134. J.M. Newsam, private communication.

135. V. Fülop, G. Borbély, H.K. Beyer, S. Ernst and J. Weitkamp, *J. Chem. Soc. Faraday Trans I* **85** (1989) 2127.

136. A.J.M. de Man, H. Küppers, R.A. van Santen, *J. Phys. Chem.* **96** (1992) 2092.

137. R. Freimann and H. Küppers, *phys. stat. sol. (a)* **123** (1991) K123.

138. J. Klinowski, *Progr. NMR Spectr.* **16** (1984) 237.

139. G.T. Kokotailo, C.A. Fyfe, G.C. Gobbi, G.J. Kennedy and C.T. DeSchutter, *J. Chem. Soc. Chem. Comm.* (1984) 1208.

140. C.A. Fyfe, H. Gies and Y. Feng, *J. Chem. Soc. Chem. Comm.* (1989) 1240.

141. J.A. Ripmeester, M.A. Desando, Y.P. Handa and J.S. Tse, *J. Chem. Soc. Chem. Comm.* (1988) 608.

142. G. Filippini, C.M. Gramaccioli, M. Simonetta and G. Suffritti, *Acta Cryst. A* **32** (1976) 259.

143. M. Könnecke, private communication.

144. G. Engelhardt and R. Radeglia, *Chem. Phys. Lett.* **108** (1984) 271.

145. J.V. Smith, *Chem. Rev.* **88** (1988) 149.

146. W.M. Meier, *Proc. 7th Int. Zeolite Conf., Tokyo* (1986) 13.

147. I. Hassan and H.D. Grundy, *Acta Cryst. B* **40** (1984) 6.

148. G.O. Brunner and W.M. Meier, *Nature* **337** (1989) 146.

149. H.J. Bosmans and K.J. Andries, *Acta Cryst. A* **46** (1990) 832. K.J. Andries and H.J. Bosmans, *Acta Cryst. A* **46** (1990) 847. K.J. Andries, *Acta Cryst. A* **46** (1990) 855.

150. B.C. Chakoumakos, R.J. Hill and G.V. Gibbs, *Am. Mineral.* **66** (1981) 1237.

151. S.W. de Leeuw, H. He and M.F. Thorpe, *Solid State Comm.* **56** (1985) 343.

152. S. Merlino, *Acta Cryst.* **A37** (1981) C189. S. Merlino, *Proc. 6th Int. Zeolite Conf., Reno* (1983) 747. G.T. Kokotailo, *Proc. 6th Int. Zeolite Conf., Reno* (1983) 760. S. Ueda and M. Koizumi, *Poster reprints 7th Int. Zeolite Conf., Tokyo* (1986) 23.

153. S. Merlino, *Europ. J. Miner* **2** (1990) 809.

154. M.J. Annen, M.E. Davis, J.B. Higgins and J.L. Schlenker, *Mat. Res. Soc. Symp. Proc. Vol. 233 (eds. R.L. Bedard, T. Bein, M.E. Davis, J. Garces,*

V.A. Maroni and G.D. Stucky) (1991) 245. M.J. Annen, M.E. Davis, J.B. Higgins and J.L. Schlenker, *J. Chem. Soc. Chem. Commun.* (1991) 1175.

155. S.L. Lawton and W.J. Rohrbaugh, *Science* **247** (1990) 1319. S.L. Lawton, private communication.

156. J.M. Bennet and B.K. Marcus, *in P.J. Grobet et al. (eds.): Innovation in Zeolite Materials Science* (1987) 269.

157. A.J.M. de Man, S. Ueda, M.J. Annen, M.E. Davis and R.A. van Santen, *Zeolites* **12** (1992) 789.

158. D.E. Akporiaye and G.D. Price, *Zeolites* **9** (1989) 321.

159. M.E. Davis, C. Saldarriaga, C. Montes, J.M. Garces and C. Crowder, *Zeolites* **8** (1988) 362. M.E. Davis, C. Saldarriaga, C. Montes, J.M. Garces and C. Crowder, *Nature* **331** (1988) 698.

160. J.W. Richardson, J.V. Smith and J.J. Pluth, *J. Phys. Chem.* **93** (1989) 8212.

161. A.J.M. de Man, E.T.C. Vogt and R.A. van Santen, *J. Phys. Chem.* **96** (1992) 10460.

162. J.V. Smith and W.J. Dytrych, *Nature* **309** (1984) 607.

163. R.A. van Santen, A.J.M. de Man, W.P.J.H. Jacobs, E.H. Teunissen and G.J. Kramer, *Catal. Lett.* **9** (1991) 273.

164. F. Schüth, *J. Phys. Chem.* **96** (1992) 7493.

165. E.H. Teunissen, C. Roetti, C. Pisani, A.J.M. de Man, A.P.J. Jansen, R. Orlando, R.A. van Santen and R. Dovesi, J. Phys. Chem., submitted.

166. E.H. Teunissen, F.B. van Duijneveldt and R.A. van Santen, *J. Phys. Chem* **96** (1992) 366.

42

Keywords
"ATOMISTIC MODELLING OF ZEOLITIC MATERIALS"

THEORETICAL CHEMISTRY IN DRUG DISCOVERY

W. Graham Richards

Physical Chemistry Laboratory, Oxford University, England

Introduction

Drugs are generally small molecules. Their mode of action is mostly based on the binding of the drug to a macromolecule. That target, or receptor, frequently has a very specific binding site and the binding free energy between drug and macromolecule is of paramount importance.

$$\text{Drug + Receptor} \quad \xrightarrow{\Delta G_{binding}} \quad \text{Drug-Receptor Complex}$$

Ideally, theoretical chemistry would furnish predictions of $\Delta G_{binding}$, prior to synthesis of possible drugs. There are economic as well as scientific imperatives behind this. Pharmaceutical companies may make several thousand compounds for every one which ever gets as far as being tested in a human being, at a cost of hundreds of millions of dollars. Not surprisingly it is a difficult problem for theoretical chemistry, but significant progress has been made.

The techniques used include quantum mechanics; molecular mechanics and statistical mechanics. Quantum chemistry with its firm physical basis would be the method of choice but for the computational demands when large numbers of atoms are involved. Molecular mechanics, the use of purely empirical potential functions, does permit the incorporation of thousands of atoms into calculations, but the major recent advances have been achieved by using molecular mechanics potentials within statistical mechanical methodologies. Most notably this has permitted the inclusion of explicit solvent molecules and counter-ions into computations which do yield free energies, or at least differences in binding free energies, not merely treating isolated molecules.[1]

The way these theoretical tools are employed depends on how much information is available at the atomic level for the problem under investigation. We can list a hierarchy of situations starting with the most tractable:

1. The receptor structure is known in atomic detail from experimental crystallography or NMR spectroscopy.

2. The receptor structure is predicted by theoretical methods.

3. The receptor remains an unknown macromolecule.

These categories of problem will now be considered separately.

Known targets

Targets for drugs which are known in some atomic detail include nucleic acids; enzymes of known crystal structure; some ion channels and macromolecules involved in the immune system.

In the case of nucleic acids the drive is to produce compounds which will bind to specified base sequences. In this way it should become possible to exploit the information coming from the human genome project by designing molecules which target and possibly switch off a given gene.

In our group in Oxford we are involved in three approaches to sequence-specific DNA ligands.

a) Spermine[2]

The polycation spermine,

$$N^+H_3 - (CH_2)_3 - N^+H_2 - (CH_2)_4 - N^+H_2 - (CH_2)_3 - NH_{3+}$$

has often been used to stabilize the polyanionic DNA. Indeed it appears in the crystal structures of most short lengths of double helix. It does, however, also have an important if not fully clarified biiological role in communicating with DNA. In the test-tube it can cause major changes in DNA structure including provoking the B-Z, right- to left-handed transformation and the enzyme responsible for its synthesis, ornithine decarboxylase, is an important target for anti-cancer drugs.

The manner of the binding to DNA is not clear. In our molecular mechanics study we suggest that it may show a preference for G-C regions of DNA. Furthermore there may be two distinct ways of binding to (poly dG-poly dC); one non-specific across the major groove, involving the negative phosphate backbone but also a highly specific down-groove mode (see Figure 1).

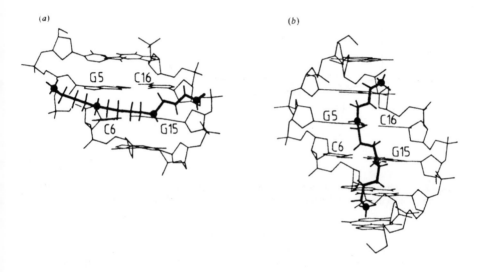

Figure 1. Minimized structures showing the complexation of spermine to the major groove of d(GCGCGCGCGC)₂ in the cross-groove (a) and down-groove (b) binding sites. The spermine molecule is represented by thickened lines with circles indicating the N atoms of protonated amine groups. Solvent molecules have been removed to improve clarity. The figure also serves to illustrate the general positions of the cross-groove and down-groove binding sites.

This down-groove mode would depend on hydrogen-bonds only possible to the O6 and N7 atoms of guanine (Figure 2).

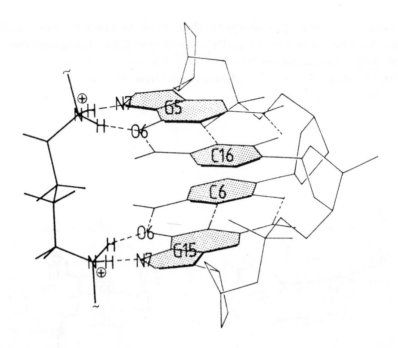

Figure 2. The proposed model of spermine binding to the poly(dG-dC) major groove. The main feature of this interaction is the efficient hydrogen bonding between spermine NH^+_2 groups and guanine N7 and O6 atoms.

If correct the structure illustrated in Figure 2 is a starting point for the design of sequence-specific ligands.

b) Organometallic compounds[3]

Molecules such as cis-platin are already very familiar as DNA ligands in their role as anti-cancer therapeutic agents. We suggest that sequence selectivity may be achieved using chiral organo-metallics. In particular ruthenium complexes of phenanthroline which have a propeller-like shape can bind differently in their lambda and delta forms. One form of binding is illustrated in Figure 3.

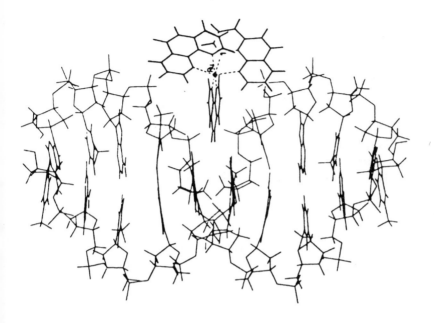

Figure 3 continued (side view)

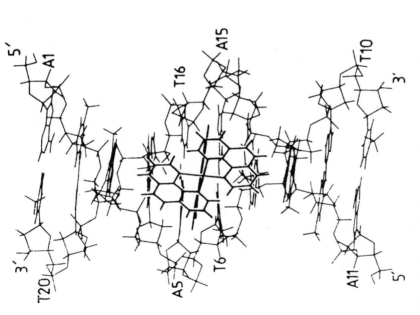

Figure 3. Energy refined structure of the d(ATATATATAT)₂ / Λ partially inserted complex

Again the point of this study is to provide a starting point for the design of lengthy sequence specific ligands.

Anti-sense compounds[4]

The use of oligonucleotides which will bind in a specific Watson-Crick manner to mRNA or to helical DNA is an industry in its own right but problems remain. It would be preferable to have non-oligonucleotide mimics. If we can firstly understand the structure of the bound antisense compound, then perhaps the free energy perturbation technique could give predictions as to how modification might influence $\Delta G_{binding}$.

This theoretical approach aims to provide differences in ΔG in the following cycle:

$$
\begin{array}{ccc}
& \Delta G_1 & \\
\text{Drug A + Receptor} & \rightarrow & \text{Drug A - Receptor Complex} \\
\downarrow \Delta G_4 & & \downarrow \Delta G_3 \\
& \Delta G_2 & \\
\text{Drug B + Receptor} & \rightarrow & \text{Drug B + Receptor Complex}
\end{array}
$$

The desired free energy difference $(\Delta G_1 - \Delta G_2)$ is equal to $(\Delta G_4 - \Delta G_3)$. These latter two free energy differences are non-physical but more readily simulated. For changes involving macromolecules the molecular dynamics approach is used to calculate free energy changes on perturbing one system to another, while for small molecules in solvent Monte Carlo techniques have advantages.

This same free energy perturbation technique has also been much applied in the area of enzyme-inhibitor studies. In our case we have been particularly interested in designing anti-cancer drugs which will not provoke the dreadful side-effects associated with contemporary chemotherapy. The basis of our strategy is the experimental fact that when tumours grow the associated blood supply from capillaries does not grow commensurately. Thus tumours get less blood and hence less oxygen than normal tissue: they are hypoxic. The notion then is to design an inhibitor of an enzyme which is crucial to cell reproduction but which can exist in oxidized and reduced forms, with the reduced form being lethal while the oxidized form is benign. At the same time the redox potential of the drug will have to be such that it will be present in the oxidized state in normal tissue but reduced in tumours.

The target chosen was dihydrofolate reductase, essential to the cell in the production of the bases for the next generation of DNA, although at the present time the more recently available structure of thymidylate synthetase is probably a better target as the cell cannot get round the blockade as easily. As a ligand, variants in methotrexate but with hydrogen binding supplied by hydroxyl groups rather than amino N-H provides a possible bioreductive drug.[5]

The free energy perturbation technique can be used not only to give binding free energy differences but also redox potentials and the all-important partition properties associated with drug transport. Redox potentials can be computed in this way to an accuracy of about 20 mV which is essentially the same level as experimental determinations.[6]

Partition coefficients are used by medicinal chemists in quantitative structure-activity studies. Drugs need to be somewhat fat-soluble in order to cross cell membranes, but not so lipid soluble that they will remain trapped. Preliminary studies showed that free energy perturbation can reproduce partition coefficients with simple solvents such as carbon tetrachloride.[7] The current research is to do similar work using a credible computer model of a cell membrane, complete with phosphatyl choline head-groups and a layer of external water. This should be very much preferable to the current experimental use of liquid n-octanol as a dubious representative of real biological membranes.

Predicted protein structures

Protein crystallography is such a lengthy procedure and many interesting targets do not form crystals. At the same time gene sequences and hence the amino acid sequence or primary structure of proteins is readily available. There is thus a massive effort to use the primary sequence to predict the three-dimensional or tertiary structure of proteins.

The most promising approach seems to be the use of homology: using the known structure of one protein to predict the shape of a related or homologous variant. We have been involved in two different aspects of homology modelling.

For cases with high homology, Gilbert[8] has shown that similarities in the properties of sequences of amino acids have advantages over considering similarities in identity. In this way Menziani et al[9] showed that there is an homology between a scorpion neurotoxin of known crystal structure and an interesting polypeptide, the so-called 'big endothelin'.

50

This latter small protein is cleaved in the body to yield endothelin, the most potent and long-lasting vasoconstrictor known. Blocking that cleavage would tend to lower blood pressure which is perhaps the largest drug market. A knowledge of the structure of big endothelin is an all-important first step.

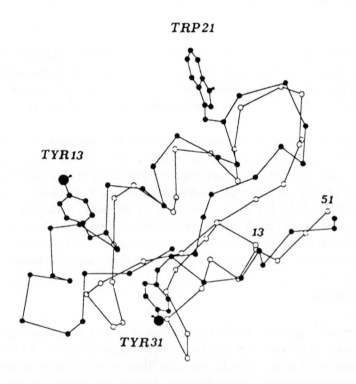

TRP21

TYR13

51

13

TYR31

Figure 4. The predicted structure of big endothelin superimposed in the scorpion neurotoxin

When homology is low these statistical methods become difficult. Morris[10] has taken a graphical approach to seeking out which bits of an unknown protein are likely to match parts of a known structure. This graphical method relies to some extent on the ability of the human eye to recognise patterns. The known and unknown protein sequences are displayed as horizontal rows of bars, each representing an amino acid with a choice of colourings available to the user. The program calculates the similarity of lengths of one amino acid sequence with similar lengths in the other. Interactively the operator can zoom in; insert gaps and finally decide on which bits of the unknown can be modelled. One useful application of this was to model an important enzyme, cytochrome P-450, used by fungi to make their cell membranes on the basis of another P-450 protein used by a bacterium to metabolise camphor.[11]

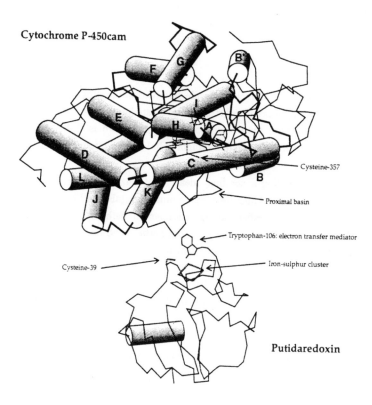

Figure 5. Putidaredoxin about to dock into the proximal basin of cytochrome P-450$_{cam}$. α-helices are shown as cylinders. The prosthetic groups and camphor are shown in broken lines. Electrons are delivered to the heme-iron at the active site of P-450$_{cam}$ from the iron-sulphur cluster via cysteine-39, tryptophan-106 and cysteine-357.

Using the predicted protein structures illustrated in Figure 5, Baldwin[12] has reinterpreted the mechanism of action of electron transport in cytochromes P-450. This detailed picture can also afford insights into how this process might be blocked by a drug or agrochemical since cytochromes represent an important class of target.

At present the prediction of the three-dimensional structure of a protein is only possible in a small range of cases, but the combination of more experimental results and the steady improvement of theoretical methods suggest that this is an area where there should be considerable progress in the next decade.

The unknown target

Knowing the structure of the macromolecule involved in the drug-receptor interaction is the best situation. Using a predicted structure is a satisfactory second-best. In the world of real problems, however, more commonly than not there is little known about the receptor, perhaps not even its molecular weight. Under these circumstances all is not lost.

One can start with Pauling's seminal idea[13] that enzymes accelerate reactions by stabilizing transition states. Thus the enzyme must recognise and bind to that transitory structure or to some intermediate along the reaction pathway. Rate acceleration is achieved by stabilizing the top of the potential surface and hence lowering the barrier between reactants and products. A stable molecule which mimicked the structure of the transition state should similarly bind to the enzyme and in competition would prevent the catalyst from speeding up the reaction. Stable transition state analogues are thus potential drugs. This is the essence of the action of penicillin which mimics the transition state of a transpeptidase.

With this background an obvious route to novel pharmaceuticals presents itself. Firstly decide which reaction one wants to block. Find the transition state. This is perhaps best done by a combination of quantum chemistry and empirical methods. Then make a stable mimic of the transition state. In order to do this, so-called 'molecular similarity' software is required.

Molecular similarity calculations aim to produce a quantitative answer to the question, 'how similar is molecule A to molecule B?' Carbo[14] introduced a measure, R_{AB}, based on the electron densities of the two molecules, p_A and p_B integrated over space and normalized so as to give a value in the range zero to unity

$$R_{AB} = \frac{\int P_A P_B \, dv}{(\int P_A^2 \, dv)^{1}/2 \, (\int P_B^2 \, dv)^{1}/2}$$

This formula is not unique and a number of others have been proposed.[15,16] For drug design work the property to be compared need not be electron density which is in fact a rather poor discriminator. Better are molecular electrostatic potential, which is related to drug binding, and shape which dominates the all-important repulsion between drug and receptor.

Until recently the electrostatic potential and shape indices were computed numerically at grid points surrounding the van der Waals surface of compounds. This has the disadvantage of being crude to the point of inaccuracy and making adjustment of relative orientations to optimize similarity prone to becoming stuck in local minima. New work[17,18] has overcome these problems by doing both the electrostatic potential and shape calculations analytically.

For MEP the value at any point in space is a sum of terms of the type q_i/r_i, with q_i being the charge on atom i and r_i the distance of the point from that same atom. The various q_i can be obtained by fitting to molecular electrostatic potentials generated by *ab initio* molecular orbital calculations. The $1/r_i$ term can be fitted by two or three gaussians. This fit is excellent beyond about half an Angstrom distance from nuclei and does not go to infinity when $r_i = 0$. Thus the calculation of R_{AB} can be performed analytically with perhaps a one hundred-fold increase in speed and far greater accuracy.

In a parallel way the shape similarity can be derived using volumes for atoms given a gaussian shape found by fitting quantum mechanical electron densities to spherical gaussians. This gives individual spherical atoms with 'soft' edges.

We have found that these measures of similarity are particularly suitable for use in quantitative structure-activity studies. In this context we use the values of similarity as input to a neural network. The approach is to take the series of compounds amongst which the activity correlation is sought and compute the similarity of every member of the series of n molecules with every other one, yielding a 2n x 2n matrix of similarity values; one for shape and one for MEP with every pair of molecules.

The columns of this matrix may then be used in a neural net such as that illustrated in Figure 6.

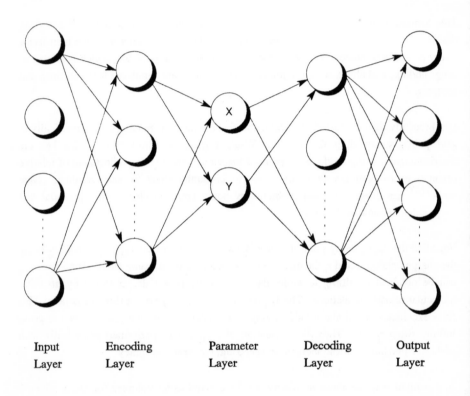

| Input | Encoding | Parameter | Decoding | Output |
| Layer | Layer | Layer | Layer | Layer |

Figure 6. The form of unsupervised neural network used for data reduction.

A network of this type takes the input and the net is trained to give output identical to the input. This is the technique of unsupervised learning which leads to dimensionality reduction. The values of the parameters x and y can be plotted against each other on a two-dimensional graph. In preliminary studies these plots have successfully separated highly active compounds from less active and inactive molecules. This can be done using a personal computer since the computational demands are not extensive.

Conclusions

Computer-aided molecular design is making a serious contribution to the discovery of new pharmaceuticals. The growing power of workstations and their falling prices bring this type of research within the capabilities of scientists who lack massive resources. Given the rewards which new drugs can bring in terms both of human health and of economic return, the future for theoretical chemists applying their talents in this area is extremely encouraging.

Acknowledgements

The author would like to thank Professor Marco Antonio Chaer Nascimento and his colleagues for the invitation to Brazil and the National Foundation for Cancer Research, pursuant to whose contract much of this work was performed.

References

1. C.A. Reynolds, P.M. King and W.G. Richards, *Molec. Phys.* **76** *(1992) 251-273.*

2. I.S. Haworth, A. Rodger and W.G. Richards, *Proc. R. Soc. Lond. B* **244** (1991) 107-116.

3. I.S. Haworth, A.H. Elcock, A. Rodger and W.G. Richards, *J. Biomolec. Struct. Dynam.* **9** *(1991) 23-44.*

4. J.S. Cohen (ed), *Oligodeoxynucleotides. Anti-sense Inhibitors of Gene Expression* (MacMillan, New York, 1989).

5. W.G. Richards and C.A. Reynolds, in *Theoretical Biochemistry and Molecular Biophysics*, eds. D.L. Beveridge and R. Lavery (Academic Press, Schenectady, 1990).

6. C.A. Reynolds, P.M. King and W.G. Richards, *Nature* **334** (1988) 80-82.

7. J.W. Essex, C.A. Reynolds and W.G. Richards, *J. Am. Chem. Soc. 114 (1992)* *3634-3639.*

8. R.G. Gilbert, *J. Molec. Graph.* **10** (1992) 112-119.

9. M.C. Menziani, M. Cocchi, P.G. De Benedetti, R.G. Gilbert, W.G. Richards, M. Zamai and V.R. Caiolfa, *J. Chim. Phys.* **88** (1991) 2687-2694.

10. G.M. Morris, *J. Molec. Graph.* **6** (1988) 135-142.

11. G.M. Morris and W.G. Richards, *Biochem. Soc. Trans.* **19** (1991) 793-795.

12. J. E. Baldwin, G.M. Morris and W.G. Richards, *Proc. R. Soc. Lond. B* **245** (1991) 43-51.

13. L. Pauling, *American Scientist* **36** (1947) 51-58.

14. R. Carbo, L. Leyda and M. Arnau, *Int. J. Quantum Chem.* **17** (1980) 1185-93.

15. C.A. Reynolds, C. Burt and W.G. Richards, *Quant. Struct. Act. Relat.* **11** (1992) 34-36.

16. A.C. Good, *J. Molec. Graph.* **10** (1992) 144-152.

17. A.C. Good, E.E. Hodgkin and W.G. Richards, *J. Chem. Inf. Comput. Sci.* **32** (1992) 188-191.

18. A.C. Good and W.G. Richards, *J. Chem. Inf. Comput. Sci.* (in press).

PROTEIN STRUCTURE DETERMINATION

SHERYL L. DORAN
BIOSYM TECHNOLOGIES, INC.
9685 SCRANTON ROAD
SAN DIEGO, CA 92121

The successful design of new drugs and the engineering of biopolymers requires an understanding of the 3D structure of a wide variety of molecules at atomic resolution. Many new primary sequences are being determined for proteins, but this has not been the case for structure determination. X-ray crystallography relies on crystallization of the protein (an often time-consuming 'art' form), and data collection and interpretation can be a long process. The alternative, NMR, provides information on the non-crystalline protein structure but has limitations on the size of molecule for which it may be used. Homology modeling Greer[1,2] relies on the fact that proteins can be classed into families according to their origin, their function and their folding patterns. Using these structural relationships, models can be built based on composite parts extracted from similar proteins, and these models can be used to give a head start on crystal structures.

The techniques of homology modeling and protein structure determination using NMR data will be discussed here.

1. Building Protein Structures by Homology

In order to build a protein model using homology modeling, certain data must be available:

- The amino acid sequence of the target protein
- The high resolution structure of at least one related (reference) protein
- Other related amino acid sequences

The more protein structures (and sequences) available, the more reliable the technique will be at providing an accurate model.

Homology modeling is based on the following assumptions:

- The 3D structure of the target (unknown) protein is similar to that of the related proteins

- Regions of homology are conserved structurally
- Biologically active residues (with related activity) have similar topology. This allows you to use active site regions (which are normally known from experiment) as regions of homology.
- Loop regions (variable sequences) allow insertions and deletions. This is now quite widely believed to be true. It is important in that it allows us to use loop regions as areas which can be used for improving sequence alignments by adding or removing residues. We assume that these changes do not corrupt the overall structure of the protein since they are normally surface-exposed and of variable size (within a family).

The overall strategy for homology model construction is:
- To find the conserved regions in the related proteins using both structural and sequence information
- To match the target (unknown) protein with conserved regions
- To assign structure to the target (unknown) protein in the conserved regions
- To complete the model in the variable (loop) regions
- To optimize the model

1.1 *Finding Conserved Regions*

The first step in constructing a protein model using homology is to perform sequence alignment on the know structures. Sequence alignment requires:
- A measure of similarity between protein sequences
- An ability to maximize the number of matches using the minimum number of insertions or deletions.

Sequences can be compared two at a time or multiply aligned.

The next step is to determine exactly what is structurally homologous among the family of known protein structures. This is done by superimposing the structures two or more at a time and observing directly where insertions and deletions occur. Structurally conserved regions (SCRs) are, by definition, those sections of the sequence which do not contain insertions and deletions. One of the most reliable alignments results from aligning an essential amino acid whose location is unambiguously known in the target sequence. Typical examples include active site residues and cysteines involved in conserved disulfide bridges. Once these single amino acids are aligned, the residues on either side (out to the extent of the SCR) are also aligned. The RMS comparisons between the proteins is monitored to determine their similarity. As an SCR is expanded, the RMS remains reasonably constant. As the SCR region ends, a dramatic increase is seen in the RMS value. In general, SCRs should not be extended over boundaries of secondary structural units. No gaps are allowed within SCRs.

1.2 *Matching Reference and Unknown Proteins*

Once all the SCRs have been defined, the process of aligning the sequence of the unknown protein to those of the known structures can begin. Sequence alignment must be used for the unknown protein since no structure is available. The new sequence can be aligned with the known sequences using either pairwise or multiple sequence alignment methods. The comparison to defined SCRs can also be used to refine the alignment.

1.3 *Building the Model*

Once the alignment of the new sequence to the SCRs of the other proteins is complete, the next step is to build a three dimensional mosaic model using coordinates from the reference proteins. To do this, the homologous protein which most closely matches the sequence of the unknown protein for each SCR is selected. Sequence scores can be determined for each protein. The backbone coordinates are extracted from these protein regions and used as a template for the structure of the unknown protein. Side chain coordinates can be extracted directly from the reference structure in cases where the residues are identical. Side chains for "mutated" residues can be generated by using the *chi* torsion values from the original residue or by using side chain conformations from rotatmer libraries, where the best steric and energetic conformation is selected.

1.4 *Loop Searching*

Until now, we have only been concerned with those residues making up the SCRs. The residues between SCRs (including insertions and deletions) we call "loops". To model a loop, search for loops in proteins which have the required number of residues and have similar structures both before and after the loop. Usually, the best source of loop templates is the homologous proteins themselves. If none of the homologous proteins have suitable loops, then searches are made of all the proteins in a structural database. Alternatively, if a reasonable loop can not be found from database searching, a loop can be created by fitting a sequence containing randomly generated *phi* and *psi* angles onto the protein in a way that does not violate steric overlap criteria with itself or the rest of the protein.

1.5 *Completing the Model*

At this point, most of the coordinates of the model have been assigned. However, the N- and C- terminal regions may not have been well defined, since they are not normally constrained to a particular secondary structure. These are normally added as linear peptides (180° dihedrals along the main chain). Also, the peptide bonds at splice points (where segments from different proteins are joined) may not be smooth. These bonds are optimized by forcing the *omega* bonds to *trans* and performing a localized minimization.

1.6 *Refining the Model*

The model may at this point may contain some unfavorable geometric or steric interactions, as it has been put together using segments from many different proteins. The final stages of refining the model consist of performing a series of minimization and dynamics calculations. These are done in stages by selectively minimizing the N- and C- termini, the variable regions or loops (side chains or all atoms), and the side chains of the SCRs (mutated and non-substituted). Finally, a series of dynamics calculations are performed with periodic minimization to give a final model which retains the structural information contained in the reference proteins along with the loop regions and mutated side chains.

2. Protein Structures from NMR Data

From specific NMR experiments, both distance and dihedral restraints can be determined for a particular protein. Distance restraints can readily be generated from NOE crosspeak volumes and dihedral restraints can be determined from ^3J coupling constants. These restraints can be imposed on the protein sequence which can be used as a starting point for distance geometry or simulated annealing (SA) calculations to determine a 3D structure for the protein.

2.1 *Distance Geometry*

The fundamental premise on which distance geometry is based is that the set of all conformations present in significant concentration can be aptly described by means of suitable distance and chirality constraints. The first are simply lower and upper bounds on the interatomic distances; the second include the handedness of the asymmetric centers in the molecule (including prochiral centers) together with the planarity of any sp^2 hybridized centers. This description, called a distance geometry description, is essentially the computer analogue of a CPK model. A large body of mathematical results exists that can be used to better understand these descriptions, much of which may be found in the recent book by Crippen and Havel[3].

Havel's distance geometry program DG-II Havel[4] is based on the following steps:

- Given the incomplete and imprecise set of distance bounds that are experimentally available, a complete and more precise set of bounds is computed via a process known as bound smoothing.
- A random guess at the values of the distances from within the bounds obtained by bound smoothing is made and atomic coordinates are found such that the distances calculated from these coordinates are a "best-fit" to this guess. Because this best-fit is computed by a projection method, this step is called embedding.
- The deviations of the coordinates from the distance bounds and the chirality constraints is minimized; this step is called optimization. By repeating the second and third steps many times with different random number generator seeds in the second step, the desired conformational ensemble is obtained. Subsequently, selected members of the ensemble of structures generated can be refined versus potential energy functions using a molecular mechanics force field.

2.2 *Simulated Annealing using a Physical Force Field*

Another method for generating families of conformations consistent with the distance and dihedral restrains derived from NMR data is to conduct a high temperature simulated annealing (SA) simulation. This SA approach as applied to a known protein sequence using approximate NMR derived distance restraints is described in Clore *et al.* [5] and Nilges *et al.* [6]. The method uses a molecular

mechanics force field and consists of searching for the global minimum of a target
function by first substantially reducing the force constants and then gradually and
selectively increasing them until they regain their full values. The target function of
which the global minimum is searched comprises the following terms:

$$F_{tot} = F_{covalent} + F_{repel} + F_{NOE} + F_{chiral} + F_{dihedral}$$

F_{tot} represents the effective potential energy in the dynamics calculation. $F_{covalent}$
describes the covalent structure of the molecule, including interresidue linkages such
as disulfide bonds, with terms for bond lengths, angles, impropers (which maintain
planarity and chirality) as defined by the selected force field. The nonbond terms of the
target function are represented by a simple repulsion term, F_{repel}. This term replaces
the van der Waals, electrostatic, and hydrogen-bonding potentials of the empirical
energy function used in molecular dynamics. Distance and dihedral restraints
derived from NMR or other physical measurements are represented by F_{NOE} or
$F_{dihedral}$. Chirality and prochirality restraints F_{chiral} are implemented by restraining
three out-of-plane angles centered at each chiral or prochiral center. The calculation
starts with an arbitrary conformation (such as a completely random array of atoms or
a linear extended chain), preventing an initial bias towards folding in any particular
direction. A sample simulated annealing protocol is described in Figure 1.

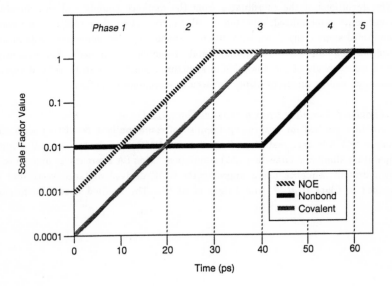

The calculations starts with minimization of the randomly distributed atoms with force constants of the covalent terms, NOE terms, and repulsive (non-bond) terms reduced to 0.001 kcal/mol/A^2. After the preparation phase, molecular dynamics with the weak force field is initiated such that the internal force constants are adjusted to values where the potential energy of the system is about equal to the kinetic energy at 1000K. The low values of the force constants in the early stages of the simulation allow energy barriers between different folds to be crossed, providing access to any region of conformation space. The force constants of the NOE restraints are scaled up to their full value during the next phase, while the covalent terms are scaled up more slowly. This loose connectivity between the atoms allows them to move fairly independently to satisfy the distance restrains, avoiding problems associated with more traditional folding methods. The repulsion force constant remains at its initial low value until the NOE restraints are at full value and the molecule is loosely folded with covalency established. The folding stage of the SA procedure is completed when all the force constants have been increased to their full values at 1000K (except the nonbond term, which is at 25% of its full value) during this regularization step. At the end of the folding phase, the covalent geometry and distance restraints should be well-satisfied. The structure is then cooled and refined in a second stage of dynamics during which the temperature is dropped to 300K and the nonbond term is increased to its full value. A final minimization completes the calculation, using a force field in which the quartic non-bond term is replaced by the Lennard-Jones form and all of the force constants are at their full values.

2.3 *Summary*

This methodology has been found to generate a high percentage (70-90%) of reasonable structures, corresponding to the provided distance restraints (Clore *et al.* [5] and Nilges *et al.* [6]). Although this method involves more computational time than distance geometry methods for small molecules, the time scales linearly with system size (assuming a constant number of total steps is required for convergence). In molecules well-defined by restraints, the structures generated are more stable in energetic terms than those obtained using other methods, particularly with respect to the nonbond contacts (Nilges *et al.* [6]). In the absence of sufficient restraints to characterize a region, this method also provides an energy-based logic for determining geometries of the molecule. SA does not suffer from the folding problems found in simple minimization and dynamics calculations, and the scaling of the constants minimizes problems at high temperatures found with these other methods.

REFERENCES:

(1) J. Greer, *J. Mol. Biol.,* **153,** 1027-1042 (1981).

(2) J. Greer, *Proteins,* **7,** 317-334 (1990).

(3) G. M. Crippen; T. F. Havel, *Distance Geometry and Molecular Conformation,* Research Studies Press, U.K. (1988).

(4) T. F. Havel, *Prog. Mol. Biol, Biophys.,* **56,** 43-78 (1991)

(5) G. M. Clore; M. Nilges; D. K. Sukumaran; A. T. Brunger; M. Karplus; A. M. Gronenborn, *EMBO J.,* **5,** 2729 (1986).

(6) M. Nigles; G. M. Clore; A. M. Groneborn, *FEBS Lett.,* **239,** 129 (1988).

Atomistic Simulation of Materials

W. A. Goddard, III, N. Karasawa, R. Donnelly, J. Wendel,* C. B. Musgrave, J-M. Langlois, K. T. Lim, S. Dasgupta, J. J. Gerdy, J-M. Langlois, T. Maekawa,@ X. Chen, H.-Q. Ding, M. N. Ringnalda,† R. Friesner,‡ T. Yamasaki,§ T. Cagin,* A. Jain,** and J. Kerins††

Materials and Molecular Simulations Center
Beckman Institute, 139-74, Contribution No. 8795
Division of Chemistry and Chemical Engineering
California Institute of Technology
Pasadena, California 91125

Abstract

Recent advances in the methodologies for atomistic modeling of structural, mechanical, electrical, and optical properties have reached the point where they can begin to play an essential role in designing advanced high technology materials. The Materials and Molecular Simulation Center (MSC) in the Beckman Institute at Caltech was established to provide the concentrated multidisciplinary efforts needed to capitalize on recent advances in atomistic modeling to build the tools required by industrial and government scientists and engineers to better design and characterize new materials.

The atomic-level methodology used here is based on first principles quantum mechanics (QM) methods in which the electronic states and structures are calculated directly. This has the advantage that no experimental information is required but the disadvantage of very restricted size and time scales. A major advance here is the pseudospectral methodology which allows *ab initio* calculations on large molecular and periodic structures.

These QM methods are used to obtain lumped parameters (referred to as a force field) to represent the electronic wavefunctions in terms of charges, spring constants, shell parameters, etc. Previously such force fields were usually experimentally based, but there have been major recent advances in the methodology (BHFF/SVD) allowing the force field to be extracted directly from QM.

With such first principles force fields we use advanced canonical molecular dynamics techniques to calculate trajectories which lead directly to thermoelastic properties (specific heat, thermal expansion, moduli) as a function of temperature. A major advance here is the Cell Multipole Method which allows practical calculations on systems having one million atoms per unit cell (needed for describing partially crystalline polymers).

The focus of this research is upon applying these new methods to predicting the mechanical, polarization, and thermochemical properties of polymers and ceramics. We are also exploring the properties at interfaces and surfaces (polymer-polymer, polymer-ceramic). Some examples of these applications are included.

*Molecular Simulations Inc., Pasadena, California 91101
†Schrödinger Inc., Pasadena, California 91106
‡Department of Chemistry, Columbia University, New York, New York 10027
§Asahi Chemical Company, Fuji City, Japan
@Asahi Glass Company, Yokohama, Japan
**Jet Propulsion Laboratory, Pasadena, California 91109
††BP America, Warrensville, Ohio

Table of Contents

1. Introduction

The advances in quantum chemistry and molecular dynamics, collectively referred to as atomistic simulations, have progressed to the point where there are enormous potential opportunities for application to numerous important industrial problems involving the materials sciences.

This opportunity is to design, characterize, and optimize materials before beginning the expensive experimental processes of synthesis, characterization, processing, assembly, and testing. With reliable simulations for real materials, industry could save enormously by cutting years off some development cycles. In addition, theory and simulation would allow more efficient consideration of completely new materials and designs. This is particularly important for the challenges of environmentally benign industrial chemistry.

The problem is that direct use of first-principles theory on the systems of most industrial interest is usually impossible. To provide the most rigorous and reliable predictions on new materials requires first-principles quantum mechanics, the solution of the Schrödinger equation to obtain electronic wavefunctions. Unfortunately the practical time and length scales for first-principles theory are often six to ten orders of magnitude too long for industrial design. Instead we envision a hierarchy of methodologies (see Figure 1) where quantum methods are the foundation. In this hierarchy each step involves averaging over the elements of deeper ones to obtain effective parameters for the next. This leads to telescoping whereby one can transcend from quantum mechanics (QM) to engineering with just a few levels of simulation.

Averaging over the electrons to obtain spring constants, discrete charges, and van der Waals parameters, QM can be replaced with Molecular Dynamics (MD) where one solves coupled Newton's equations to predict the motions of systems 100 times larger for periods about 10^3 times longer. Similarly to obtain macroscopic properties one averages over the atoms to obtain continuous parameters for use in coarse grain modeling (finite element calculations, continuum mechanics, Navier-Stokes equations, Maxwell's equations). Such continuum studies are the meeting ground between first-principles theory and methods of engineering design.

Currently we concentrate on the first two level in Figure 1:

i. quantum mechanics

ii. atomic level molecular dynamics

and the first two linking connections

iii. averaging the electronic degrees of freedom into force fields and charges

iv. lumping of atomic degrees of freedom into macroscopic parameters.

Despite tremendous progress in the methods for QM and MD simulations over the last few years, the sizes and times required are far too short for real materials science problems. The MSC research projects are designed to address these issues over the next few years by focusing on:

a. developing methods of quantum mechanics and molecular dynamics that *(i)* scale linearly with size, *(ii)* are orders of magnitude faster than current methods, and *(iii)* are sufficiently localized for application of scalable massively parallel computers,

b. developing optimized computer software implementing these methods on scalable massively parallel or distributed computers,

c. applying these methods to the real materials of interest to industrial research and development, and

d. transferring the technology to industrial research and development teams by intense collaborations in which the industrial scientists and engineers work shoulder to shoulder with the theorists developing the new methods.

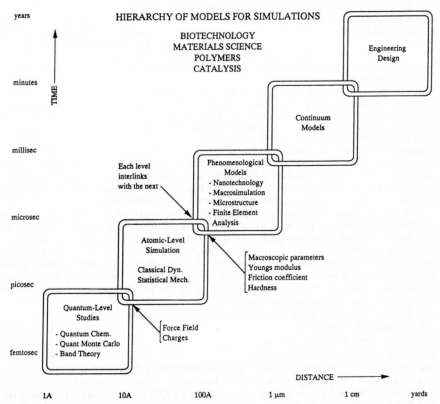

Figure 1.

The recent progress in theoretical methodologies combined with the continuing progress in scalable massively parallel hardware convinces us that the bottlenecks to developing theory and simulation tools for design and characterization of real materials can be solved in the next few years.

However developing the methods and software is only half the job. Most advances using theory to help solve real industrial problems have had *finesse* as a critical component. The scientist with a thorough understanding of both the basic theory and the application simplifies the theory to make the calculations practical while ensuring that the simulation still adequately describes the application of interest. As a result, progress is most effectively made by coupling the theorists at the cutting edge of theory and simulation with the chemists, biotechnologists, or material scientists driven to design new or improved materials, catalysts, processes, etc.

There is a difficulty in achieving this coupling. The basic theoretical research is concentrated in universities, and these groups often have little direct knowledge of the most important industrial applications. In addition, the theoretical efforts tend to be fragmented with no central location where a variety of theoretical efforts are coordinated to focus on providing the breakthroughs needed for real applications. On the other hand, even the largest industrial organization usually has at most only a handful of theorists, far short of the critical mass needed to keep up with the research advances in the various fundamental areas of theory or with the applications to various fields.

To help achieve these goals we have established as part of the Beckman Institute at Caltech, the Materials and Molecular Simulation Center (MSC) with two major goals:

1. Focus theoretical research on the key bottlenecks that obstruct application to the important materials science problems.
2. Facilitate the technology transfer of advances in materials and molecular simulations from university research laboratories to industrial practice.

The overall objectives of the MSC are to:

a. develop the tools necessary for microscopic (atomic level) modeling and simulation of the mechanical, physical or chemical properties of aerospace materials
b. validate the tools by application to properties and systems amenable to experimental verification
c. illustrate the strategies and techniques in using these tools by applying them to relevant industrial problems
d. initiate the development of specific models by tackling the applied problems of general importance to industry or of specific interest to industral collaborators
e. develop solutions to critical bottlenecks for applications to advanced materials
f. transfer this technology to industry through collaboration and workshops
g. assist industrial and government laboratories with support and guidance on theory, modeling, and simulation.

The sources of the funding so important in establishing the MSC are summarized in Section 8.

2. Collaborations
2.1 Industrial Collaborations

We have established effective collaborations or interactions with a number of industries including: Allied-Signal, Asahi Chemical, Asahi Glass, BP America, BF Goodrich, Chevron, General Electric, General Motors, Gore Corporation, Molecular Simualtions, Inc., Schrödinger Inc., and Xerox. These interactions all focus on applications of atomistic modeling to real materials. Current collaborations include:

1. Force fields and properties of nylon type polymers (Willis Hammond and Daryl Boudreaux of Allied-Signal, Morristown, New Jersey).
2. Force fields and properties of silicon oxycarbide ceramics (Charlie Parker and Steven Griener of Allied-Signal, Des Plaines, Illinois).

3. Properties of polyethylene and of acrylonitrile-butadiene copolymers and blends (John Kerins and Jim Burrington, BP America, Warrensville, Ohio).
4. Glass properties and moduli of polystyrene-maleic anhydride copolymers (Yongchun Tang, Chevron Petroleum Technology, La Habra, California).
5. Force fields and properties of oxygen containing polymers (Ken Smith and Randy Stewart of General Electric Research and Development, Schenectady, New York, Tahir F. Cagin, Feng Fan of Materials Simulation Inc., Pasadena, California).
6. CVD growth of diamond film (Steve Harris of General Motors Research, Warren, Michigan).
7. Atomic level studies of nanotechnology (Ralph Merkle of Xerox, Palo Alto, California).
8. Glass temperatures and moduli of chlorinated polyvinylchloride polymers (R. Wissinger, D. White, P. Adriani, and J. Schnitzer of BF Goodrich, Brecksville, Ohio).
9. Gibbs canonical and grand canonical dynamics (Tahir F. Cagin and Zhou-Minh Chen, Molecular Simulations Inc., Pasadena, California).
10. Advanced pseudospectral methods for electronic wavefunctions (M. Ringnalda, R. Donnelly, Schrödinger Inc., Pasadena, California).
11. Liquid crystals (Takashi Miyajima of Asahi Glass).
12. Optimization of accurate force fields (T. Yamasaki of Asahi Chemical).

These collaborations have been set up so that one scientist at the MSC and one scientist at the industrial laboratory is assigned to each collaboration. The industrial scientists generally spend between 2 and 52 weeks a year at the MSC working with the MSC scientist on the collaboration. Between such visits are frequent other contacts. In this way the industrial scientist becomes thoroughly familiar with the strategies and techniques used to finesse a particular applications, rather than just seeing the final answer. The MSC shares computer programs with the collaborators and allows them to remotely use the MSC facilities. As the collaboration is successful and the industrial collaborators become facile with the simulation tools, the industrial laboratories apply these tools to other materials science problems at their laboratories.

The time scale for most collaborations is 2 to 4 years and the specific projects change as the simulation technology is developed and validated. These collaborations provide a strong driving force toward finding ways to make the simulations suitable for applications to real materials and ensure that the MSC

 $i.$ helps solve the materials science problems of most interest to industry and

 $ii.$ effectively transfers the technology to industrial research and development laboratories.

2.2 University Collaborations

Critical to the MSC project is developing the new methods and software required for high capacity simulations of real materials. MSC has established collaborations with leading outside groups including:

 $a.$ Richard Friesner, Professor of Chemistry and Director of the NIH Biophysics Resource Center, Columbia University. This collaboration involves pseudospectral methodology.

 $b.$ Dr. Richard P. Messmer, Adjunct Professor of Physics, University of Pennsylvania, Philadelphia, and Coolidge Fellow, GE R&D, Schnectady, New York. Dr. Messmer focuses on new force fields and simulations of metal alloys.

c. Zhen-Gang Wang, Assistant Professor of Chemical Engineering, Caltech. Professor Wang focuses on self-assembly of amphiphilic molecules (micellized surfactant solutions, mesoscale structures of diblock copolymers) with an emphasis on mechanical response of these materials on short time scales.

d. Stephen Taylor, Assistant Professor of Computer Science, at Caltech. Professor Taylor is the co-developer of PCN (program composition notation) for automatic development of optimized software for scalable massive parallel computers. The collaboration focuses on developing refined automated tools for quantum and molecular dynamics simulations on highly parallel computers.

e. Abhinandan Jain and Guillermo Rodriquez research engineers in Robotics at the Jet Propulsion Laboratory at Caltech. In developing software for control of spacecraft robots, they developed the basic NEIMO approach for internal coordinate dynamics. We are collaborating on developing practical programs for molecular dynamics on polymers and materials systems and implementing NEIMO on parallel computers.

f. Professor Tony Rappè, Professor of Chemistry, Colorado State University, Fort Collins, Colorado. Professor Rappè focuses on developing new generic force fields suitable for wide classes of materials.

g. We also provide opportunities for experts in theory or computers to spend 6 to 12 month sabbaticals at the MSC interacting with the university and industrial scientists and enginners on site.

3. Developments in Atomistic Methodologies

In this section we summarize some of the recent advances in theoretical methodologies that serve as the basis for the thrust into simulations of real materials.

3.1 Pseudospectral Quantum Chemistry

A major focus of the Goddard research group in the 1970's and 1980's, was to develop methodology and procedures for *ab initio* quantum chemical studies on clusters to mimic the chemistry at active sites for homogeneous catalysis, heterogeneous catalysts, surface reconstruction, excited states of molecules, etc. This research led to the GVB suite of programs[1,2] which provides powerful methods of including electroncorrection (many-body) effects for accurate predictions of reaction intermediates in transition metal complexes. These studies led to a number of insights on the chemistry at transition metal centers[3] and led to new or modified mechanisms for key steps of several important catalytic reactions.[4,5,6]

Despite the progress, there remain severe limitations in these previous studies. *Ab initio* methods with a high level of electron correlation are essential to obtain accurate predictions about reaction intermediates and saddle points of reactions (since there is little experimental data to use as guidelines in adjusting parameters). However, this restricts us to using small clusters as models of catalytic systems, making it difficult to examine the role of promoters, poisons, modifiers, solvents, and surface structures on catalytic pathways and rates.

There are two major difficulties facing the use of *ab initio* methods on real materials. First, the methods scale badly with size (as N^4 for GVB and HF and N^5 or worse for higher levels of electron correlation). Second these techniques are difficult to incorporate with periodic boundary conditions. Methods for solving both of these problems have recently been developed. These advances start with the Pseudospectral (PS) method developed by Professor Richard Friesner[7-9] of Columbia University. In PS most of *the two-electron integrals are never calculated.* Instead, the Coulomb and exchange *operators* required for the Fock operators are *evaluated directly over a numerical grid.* With appropriate dealiasing and multigrid strategies, the PS method achieves the accuracy of standard *ab initio* methods while scaling as N^3 ($N^{2.2}$ with cutoffs) rather than N^4. The Goddard and Friesner groups have collaborated over the last three years to incorporate the PS technology into optimized software for including electron correlation[10] (PS-GVB), many-body perturbation theory (MP2) and periodic boundary conditions[11] (PS-GVB). The PS-GVB code[12] now allows practical calculations with 2000 basis pairs (100 to 200 atoms with good basis sets and effective core potentials[13,14]), and we are now beginning to implement PS-GVB on massively parallel computers (see Section 4.1).

Even more important, PS greatly simplifies the methodology for using Periodic Boundary Conditions (PBC) in HF and GVB calculations.[11] The reason is that the real space Coulomb and exchange operators are evaluated quite simply in PS. This reduces the computational work by a factor of N^2 with respect to standard *ab initio* methods, making PBC practical. In addition the use of grids allowed full implementation of point group and space group symmetry. We have developed from scratch a completely new PS-GVB-PBC program implementing PBC into HF and GVB wavefunctions.[11] This project has taken over two years and the scalar nonparallel version should be completed in 1993.

In developing PS-GVB-PBC, the PS method was extended to deal with long range basis functions in a crystalline environment. This involves using an Ewald-like method to take care of the long range Coulomb potential.[11] In standard Ewald methods[15] the sum over the Coulomb potential is partitioned into two pieces, where part of the summation is done in real space and the other in reciprocal space. Both summations converge rapidly when an appropriate screening function is added to the summations in reciprocal space and subtracted from the summations in real space. For PS-PBC, the Ewald method was modified in order to deal with the continuous nature of electron distribution for Gaussian orbital bases. The resulting Gaussian Ewald method[11] involves a different screening function for each pair of Gaussian type orbitals that cancels exactly the long range behavior of the Coulomb potential in real space.[11] Special care was also used to deal with the $k = 0$ contribution in reciprocal space which defines the absolute reference of energy.[11] The reciprocal space summation is done analytically (the Coulomb potential has a simple form in reciprocal space). The real space summation is done using PS. In PS, the action of the Coulomb potential, J_g, on a localized atomic orbital, $\chi_{\mu g}$, is evaluated directly over a set of grid points, where g is the grid index and μ labels an atomic basis function. This function is then numerically integrated over the grid with a least-squares matrix, $Q_{\mu g}$, which depends on the atomic orbital μ. The final Coulomb operator for the $\mu\nu$ pair of atomic orbitals becomes

$$F_{\mu\nu} = \sum_g Q_{\mu g} J_g \chi_{\nu g}, \qquad (3.1-1)$$

where the sum over grid points is restricted to the region of space where $Q_{\mu g}$ and

$\chi_{\nu g}$ have non-zero overlap. Thus the PS method involves two steps: *(i)* evaluation of the fitting matrix (Q) for a given geometry and basis set and *(ii)* evaluation of the Coulomb field on the grid to each step of the self-consistent part of the calculation. For crystalline systems with small unit cells, all Coulomb operators involving long range (LR) basis functions are done in reciprocal space. The only contributions that need be done in real space using PS are the ones involving short range (SR) basis functions in both the field (J_g) and in the atomic orbitals $(Q_{\nu g}$ and $\chi_{\mu g})$. For such systems the construction of the fitting matrix is simple since the Q matrix for SR functions need only include basis/dealiasing functions from nearby atoms.

For large unit cells there are some contributions from LR functions that must be done in real space. Here one needs also to construct a fitting matrix for LR functions. In the standard PS method, the cost of constructing Q for LR functions becomes prohibitive since too many basis and dealiasing functions must be included due to the LR nature of the basis functions. We find that the efficient way around this problem is to separate the different length scales present in the calculation. Thus the fitting matrix for LR functions is designed only to numerically integrate fields which are also LR in nature, thereby reducing the construction of the Q_{LR} matrix to a managable task since many fewer basis/dealiasing functions are now used. For the exchange operators it is not possible to use an Ewald-like method to split the sum into real space and reciprocal space summations. In this case, we use the PS method to assemble the exchange operators in real space. Here efficient cutoffs are used to assure rapid convergence. The convergence is fastest for insulators with very open structures (e.g. C_{60} crystal). In such cases, the density matrix falls off exponentially, allowing an efficient summation of the exchange potential. Another necessary improvement for crystalline systems was the introduction of symmetry into the PS method.[11] In the self-consistent part of PS calculations, symmetry is easily implemented by restricting the grid points to a symmetry wedge of the first unit cell (e.g. 1/48th for an O_h space group). The Coulomb and exchange operators are first obtained over the wedge. Then all the space group operations are applied to these reduced operators to obtain the full operators, allowing PS to take full advantage of crystal symmetry.

For accurate prediction of electronic band gaps and other crystalline properties, one needs to include electron-correlation effects (GVB) missing in HF theory. General use of PBC for GVB wavefunctions is complicated because the individual GVB Fock operators do not combine into a single effective multiparticle operator.[1] However for completely filled bands, GVB-PP wavefunctions are straightforward for PBC. This requires additional localized exchange terms but these can be calculated using PS. The restriction to closed shells (filled bands) is no limitation for most problems involving polymers, semiconductors and ceramics, although it precludes study of metallic systems (with partially filled bands).

Such GVB-PBC calculations of three-dimensional crystals should be very valuable in extracting charges and force fields that could be used for molecular mechanics and molecular dynamics calculations of noncrystalline systems. By using finite thickness slabs and PBC, we can study crystal surfaces, allowing us to consider structures and energetics for various reaction intermediates.

3.2 Pseudospectral Methodology and Local Density Functionals

The Goddard group has only recently been involved with developing or using

Local Density Functional (LDF) methodology for QM calculations. However, recent advances in LDF theory convince us that LDF is an essential tool in applications of theory and simulation to real materials. Thus Andzelm and Wimmer[16] have shown that with gradient corrections LDF gives excellent accuracy for ground state properties of molecules while Chen and Toigo[17] have shown a simple way to include nonlocal corrections. Most important is the recent breakthrough by Baroni and Giannozzi[18] who suggest an approach (using LDF) whereby the energy and density can be obtained from calculations that scale linearly with the size of the system! The new developments in QM methods at the MSC combine *ab initio* PS methods with LDF methods, using similar grid and antialiasing strategies for both. LDF has the same scaling properties as PS-HF and PS-GVB ($N^{2.2}$ with cutoffs), and we believe that the PS multigrid and antialiasing strategies will allow increased accuracy with smaller grids (lower costs) for LDF. Thus the methodology of the PS-HF-PBC program is suitable for LDF and we are developing a companion PS-LDF program.[20] Of particular interest here are Car-Parrinello[19] type calculations in which the molecular dynamics uses forces directly from quantum chemistry (rather than force fields). The PS technology should allow this to be practical for both *ab initio* and LDF wavefunctions.[20]

3.3 Cell Multipole Methods

Real polymer materials are often partially crystalline and simulations should involve a number of crystallites (each involving say, 20,000 atoms) interspersed with connecting loops and amorphous regions. Consequently, simulations of real polymers may require 100,000 to 1,000,000 atoms per unit cell. This requires major improvements in methodologies since current simulations are generally on systems 10^3 times smaller.

The biggest bottleneck obstructing atomic-level simulations on super-large systems is accurately summing the Coulomb interactions, which decrease slowly with distance and could lead to $N^2/2 = 0.5 \ 10^{12}$ terms for a million particle system. The standard approach to simplifying such calculations for finite systems has been to use nonbond cutoffs with spline smoothing.[21,22] However this leads to an enormous nonbond list for one million particles and also leads to errors two orders of magnitude too large. The only reliable previous procedure (Ewald) for summing the Coulomb interactions for a periodic system requires Fourier transforms[15] which scale as N^2 or $N^{1.5}$, totally impractical for one million atoms.

3.3.1 The Basic Cell Multipole Method[23]

The key steps in CMM are as follows[23]:

1. Divide space into uniform cells. A box is placed around the full molecule (about 250Å on a side for one million particles) and divided into eight equal sized cubic children cells, each of which is divided into eight grandchildren cells, etc., until at the lowest level there are about four particles per cell. For one million particles, this leads to levels 0 to 6, as indicated in Figure 2.

2. Compute multipole moments for each cell. These moments are the charge Z; dipoles (μ_α); quadrupoles ($Q_{\alpha\beta}$), etc., where it is sufficient to stop at quadrupole or octupole. The moments for the microcells (level 6 for one million particles) are shifted and added to obtain the moments for the parents, the parents are shifted and added to get the moments of the grandparents, etc.

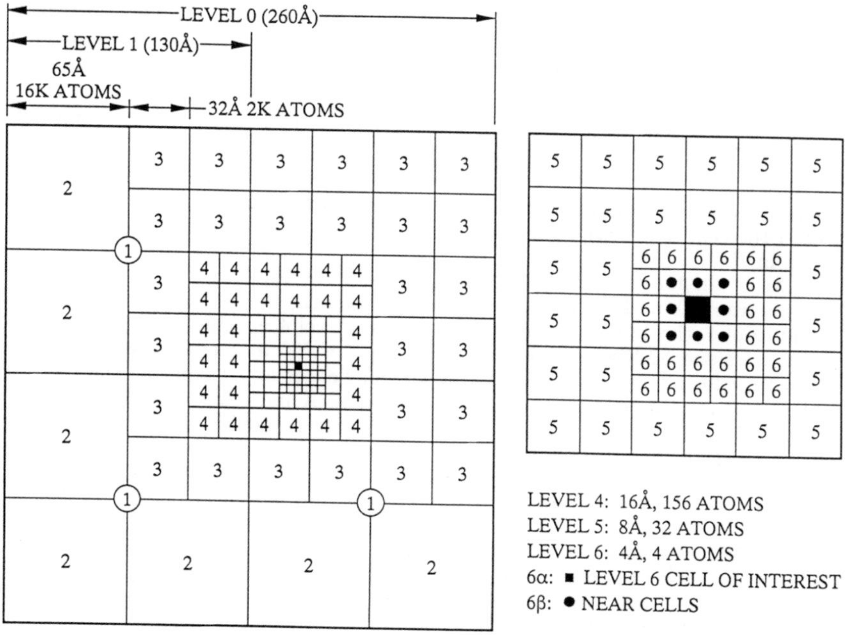

Figure 2. Cell multipole hierarchy for a one million atom system.

3. Use the multipole series (expanded about the center of the cell A),

$$V_A^{pole}(R) = \frac{Z}{R} + \sum_\alpha \frac{\mu_\alpha R_\alpha}{R^2} + \sum_{\alpha\beta} \frac{Q_{\alpha\beta} R_\alpha R_\beta}{R^4} + ... \qquad (3.3-2)$$

to describe the interactions of the atoms in each cell A with all other atoms

4. Partition the interactions in terms of near-fields and far-fields. Consider the atoms in a particular cell 6α of level 6, Figure 2. The interaction with atoms which are in the same cell or one of the $3^3 - 1 = 26$ neighbor cells (denoted 6β in Figure 1) are calculated exactly. We refer to these 27 cells as the near cell for atom i and refer to all other cells as far cells. The total potential is thus decomposed as

$$V(R_i) = V_{far}(R_i) + V_{near}(R_i), \qquad (3.3-3)$$

where

$$V_{near}(R_i) = \sum_{j,near} \frac{Q_j}{R_{ij}} \qquad (3.3-4)$$

$$V_{far}(R_i) = \sum_{A,far} V_A^{pole}(R - R_A) \qquad (3.3-5)$$

5. Convert multipole fields to Taylor coefficients. To calculate $V_{far}(R_i)$ efficiently, we avoid repeating the sum over all far cells for each atom i in a cell by doing a local Taylor series expansion about the center of the cell

$$\sum_A V_A^{pole}(R - R_A) = V^{(0)} + \sum_\alpha V_\alpha^{(1)}\tau_\alpha + \sum_{\alpha\beta} V_{\alpha\beta}^{(2)}\tau_\alpha\tau_\beta + ... \qquad (3.3-6)$$

Here both the atom position R and the cell position R_A are with respect to the center of the cell. The Taylor coefficients are computed by expanding each multipole term in R.

6. Construct the far field. Because of the hierarchy, the total far field (due to all one million atoms but the ~108 nearby ones) is constructed by adding the 189 level 6 multipoles, the 189 level 5 multipoles the 189 level 4 multipoles, the 189 level 3 multipoles and 39 level 2 multipoles for a total of 795 terms (see Figure 2). The total time for all aspects of calculating the far field is about equal to the time to calculate the near field (with ~54 interactions). A particular advantage of this simple algorithm is that the far field portion (represented by the Taylor series) changes very little during dynamics or minimization. Thus for many systems one can use the same Taylor coefficients (without updating) for 100 dynamical time steps, reducing the computation to the cost of considering just the ~54 nearby atoms!

The CMM method has been implemented on the Kendall Square Research (KSR) parallel supercomputer (64 nodes, 2 GByte memory, 20 GByte disk) newly installed in the MSC. For 1.2 million atoms the elapsed time per molecular dynamics step is now 54 sec and we expect improvement as the software is optimized.

Figure 3 shows the results of applying CMM to starburst dendrimer polymers.[24] The larger calculations considered clusters of generation 9 PAMAM (poly β alanine) dendrimers[25] each with 15333 atoms. Up to 79 of these were clustered together; leading to calculations of up to 1,211,307 atoms. Figure 3 shows that the computations indeed scale linear with size. Results are shown for two different computers, one involving a single processor Silicon Graphics 4D/35 computer with all results kept in real memory. The second used a new parallel supercomputer from Kendall Square Research (KSR). The KSR used 30 nodes with a total of 1 gigabyte fast memory and 10 gigabytes of disk. The special virtue of KSR is a memory cache system that greatly simplified writing of codes for parallel environments (the programmer need not worry about message passing). The results shown used the CMM program developed in just two weeks! The computation time for the Coulomb interactions for this 1.2 million atom system is now just 10 sec (for the 30 node KSR).[26]

In addition to computer time, a second issue with CMM is accuracy. The standard method currently used for large molecules is to smoothly cutoff the interactions between R_{in} and R_{out} using a cubic spline function.[21,22] We compare in Figure 4 the use of spline cutoffs and CMM for various choices of cutoff parameters.[27] Using only quadrupole terms in CMM leads to an RMS error of 0.2 (kcal/mol)/Å while including through octupole terms leads to an error of 0.1 (kcal/mol)/Å. In contrast standard values of spline cutoffs[21,22] lead to errors of about 19 (kcal/mol)/Å.[27] To understand the significance of such errors, we find that the RMS coordinate error for the structure optimized with CMM is less than 0.06Å whereas the spline cutoff leads to an RMS error of about 0.8Å.

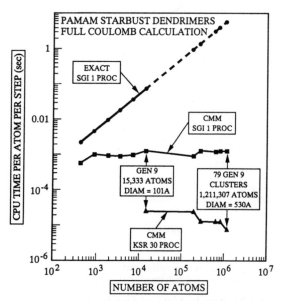

Figure 3. Computational times per atom for starburst dendrimers with up to 1.2 million atoms. This includes both the near field and far field Coulomb terms. Tests were carried out using a single scalar processor on the Silicon Graphics 4D/380 workstation and with 30 nodes of the Kendall Square Research Computer.

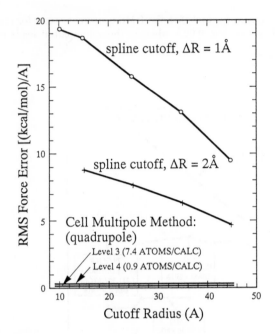

Figure 4. Comparison of the accuracy of spline cutoff methods with CMM. The test system is the generation 7 dendrimer with 3844 atoms.

3.3.2 The Reduced Cell Multipole Method (RCMM)[28]

To obtain accurate energies for Coulomb potentials in infinite systems requires the Ewald method,[15] which separates the Coulomb sum into two parts, one of which converges rapidly in real space, and the other of which converges rapidly in reciprocal space (Fourier transforms). However, the calculation time for Ewald involves terms scaling like N^2, making it impractical for large systems. Optimizing the cutoffs[15] reduces the scaling to $T \sim N^{1.5}$, but Ewald remains impractical for crystals with a million atoms per unit cell.

We have developed a new and efficient approach, the Reduced Cell Multipole Method[28] (RCMM) to compute the Coulomb and van der Waals interactions for crystals. The main elements are:

1. use a Reduced Cell of P=35 atoms to replace all distant unit cells;
2. compute the interactions of the 26 neighboring million-atom cells using the CMM.

The standard Ewald procedure[15] is applied only to the Reduced Cell so that the calculation time becomes a fixed constant independent of the system (for one million atoms this takes 2 sec on a SGI 4D/35 workstation).

The central unit cell plus its 26 neighbor cells are referred to as neighbor cells. The remaining 27 cells are referred to as distant cells. In each distant cell, we replace the original 1,000,000 atoms with a Reduced Cell of 35 atoms which has the same multipole moments up through dodecapole. The principle behind this Reduced Cell approach is that the difference between the Reduced Cell system and the original million-atom cell involves only high order multipoles and is both small and absolutely convergent. All subtleties associated with the conditional convergence of Coulomb sums are taken care of by the P atoms in the Reduced Cell, which is evaluated with the standard Ewald method.[15] Thus for any atom i in the central cell, the potential is written as

$$V^N(R_i) = V^{N-P}(R_i) + V^P_{EW}(R_i) \qquad (3.3-7)$$

where V^N is the exact potential due to the original cell of N atoms, V^P is due to the Reduced Cell of P atoms, and V^{N-P} is due to the combined $N+P$ atom system where the P charges of the reduced set have their signs reversed. The subscript EW denotes that the potential is calculated using the Ewald procedure. V^{N-P} is absolutely convergent since the first nonzero multipole moment is of order 5 which falls off very quickly $(1/R^6)$. Thus the Ewald procedure is not needed in computing V^{N-P}. Indeed, V^{N-P} can be well approximated by including only the interactions from neighbor cells. Therefore, we have

$$V^N(R_i) \cong V^{N-P}_{nbr}(R_i) + V^P_{EW}(R_i) = V^N_{nbr}(R_i) + V^P_{dist}(R_i) \qquad (3.3-8)$$

where

$$V^P_{dist}(R_i) = V^P_{EW}(R_i) - V^P_{nbr}(R_i) \qquad (3.3-9)$$

and nbr indicates the 27 neighbor cells. The subtraction of V^P_{nbr} in (9) is conveniently included in the real space sum inside the Ewald calculation of V^P_{EW}. We are left with calculations of a finite system composed from the 27 neighbor unit cells.

3.3.3 Application to Million Atom Crystals[28]

Figure 5 shows the result of using the Reduced Cell Multipole Method (RCMM) to compute the energy and forces on periodic amorphous polyethylene with unit cells containing up to one million atoms. The total computational time per atom, T/N, is constant. For the 1025536 atom/unit cell polymer, RCMM uses a total of 0.90hr, (which becomes 0.46hr per time step), whereas Ewald would require 642hr and Minimum Image would require 3368hr. The memory requirements for RCMM are modest. Storing multipoles, Taylor coefficients, and cell indices requires 120M+8N bytes, where M is the total number of boxes in a unit cell. Including 28N for coordinates, charges and forces, and assuming M≈N/4 (optimum choice), the total memory is ~70N bytes.

The RCMM is well suited to parallel or distributed computers and to vector processors. The Ewald calculation for the $V^P_{dist}(\mathbf{R})$ due to the Reduced Cell is an insignificant computation (2 sec on a workstation) and can be computed on a single processor of the parallel computer or on the scalar mode of a vector computer. The remaining 27 cells containing the original N atoms is computed using the CMM, which is dominated by the deeper level calculations. This is highly localized and therefore well suited to parallel computers (see Section 4.2). The loop structure using Cartesian coordinates makes it easily vectorized for vector computers.

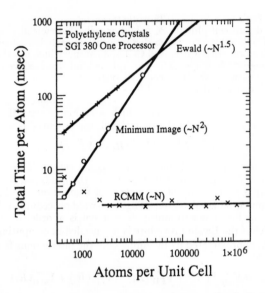

Figure 5. Computational times per atom for amorphous polyethylene crystals with up to 1 million atoms per cell. All tests were carried out using scalar processor on the Silicon Graphics 4D/380 workstation.

3.4 Combined QM/MD

For such problems as transition states of chemical reactions, it is very difficult to determine a force field sufficiently accurate for the simulations (however, see Section 3.6). The alternative is to simultaneously do both quantum chemistry and molecular dynamics, denoted as QC/MD. For each configuration of the nuclei one carries out the QC calculation on the electrons (e.g., HF, GVB) to obtain the forces on each atom and then uses these forces to increment the positions of the nuclei (MD). The first such formulation was due to Car and Parrinello[19] who used a simplified density functional theory for the QC part. We are implementing on the KSR and Intel parallel super-computers the PS-GVB, PS-LDF, and MD/CMM simulations and will then develop a QC/MD program. With PS we expect the costs of exact exchange and local density calculations to be comparable. The major theoretical issue is how to handle the time dependence of the kinetic energy terms arising from the MD momentum conjugate to the electronic wavefunction. Recent developments by B. Berne[29] suggest how to use the Louiville equation to optimize iterations combining the different time scales of electrons and nuclei.

3.5 Averaging over Electronic Degrees of Freedom (Force Fields)

Despite the progress in treating larger clusters with *ab initio* quantum chemistry, the calculations remain far too slow for studying the *dynamics* of realistic atomistic models of polymers, ceramics, and composites. Thus it is essential to replace the electrons with a force field suitable for molecular dynamics and Monte Carlo simulations. Typically force fields are described in terms of short-range valence (bonded) interactions (E_{val}) expressed in terms of two-body (E_{bond}), three-body (E_{angle}), and four-body ($E_{torsion}$, $E_{inversion}$) terms,

$$E_{val} = E_{bond} + E_{angle} + E_{torsion} + E_{inversion} \qquad (3.5-1)$$

and long-range nonbonded interactions (E_{nb}) composed of van der Waals (E_{vdw}) and electrostatic (E_Q) terms

$$E_{nb} = E_{vdw} + E_Q \qquad (3.5-2)$$

so that the total potential energy is

$$E = E_{val} + E_{nb} \qquad (3.5-3)$$

We have made significant progress in describing each of these quantities with a general scheme suitable for polymers, ceramics, and semiconductors.

3.5.1 Electrostatics

Given a set of charges Q_i on various atoms, the electrostatic interaction is

$$E_Q = \sum_{i>j} \frac{Q_i Q_j}{\epsilon R_{ij}} \qquad (3.5-4)$$

3.5.1a Potential Derived Charges[30] (PDQ)

Traditionally, the charges for molecular simulations have been estimated on the basis of quantum chemistry calculations. The simplest procedure used Mulliken populations [where the molecular orbital (MO) coefficients are used to obtain atomic populations]. This does not lead to accurate electrostatic forces and most workers now base molecular charges on the *electrostatic field* from *ab initio* Hartree-Fock (HF) calculations using a good basis set (double zeta polarization, e.g., 6-31G**). In this procedure (called *potential derived charges or PDQ*), the electrostatic field $V(R_g)$ at a grid of points R_g, is calculated using the density, $\rho(r_e)$, from the HF or GVB wavefunction.

$$V(r_g) = \int d^3 r_e \frac{\rho(r_e)}{R_{ge}} \qquad (3.5-5a)$$

$$\rho(r_e) = \sum_i 2f_i |\phi_i(r_e)|^2 \qquad (3.5-5b)$$

where f_i is the occupation number (1 for closed shell orbitals). Using the grid points outside of the vdw radii, atom-centered charges are derived so as to match the HF potential.

We have modified the PS-GVB program (the PS-Q module) to automatically use this procedure for evaluating charges from HF and GVB-PP wavefunctions.[30] Currently

PS-Q allows up to 2000 basis functions (with a good basis set this can handle molecules with up to 100 C + 100 H atoms) and we now routinely use such PDQ charges for polymer and ceramic clusters.

3.5.1b Charge Equilibration (QEq)[31,32]

There are several serious difficulties with the PDQ approach to determine the charges for simulation of materials.

    a. **Size**: Sufficiently accurate HF calculations (631G** basis) are practical only for smaller molecules (~100 atoms) and the calculations to obtain the wavefunction may take too long.

    b. **Polarization**: The charges should be allowed to polarize in response to external electric force and upon adding ions or polar molecules.

    c. **Dynamics**: For some problems (e.g. metallic systems, chemical reactions), we would like to allow the charges to respond instantaneously as atoms move about and as bonds are broken. In order to recalculate the charges during the dynamics, the charge calculation must be extremely rapid.

    d. **Crystals**: Many problems of interest involve crystals. Here there are two problems. First it is only now becoming possible to calculate the HF or GVB wavefunctions for the crystals. Second there is no outside for evaluating the potentials due to the charges.

What is needed is a general approach that can provide charges for large or infinite systems while allowing the charges to polarize and change during the dynamics.

The *Charge Equilibration* (QEq) method[31,32] provides, we believe, a solution to these problems. The basic idea is to consider each atom as a separate spherical system in a grand canonical ensemble. The chemical potential for each atom A is calculated as

$$\chi_A = \frac{\partial E_Q}{\partial Q_A} \tag{3.5-6}$$

and charge is allowed to flow until the chemical potentials are equal. The charge energy E_Q used to calculate the chemical potential is written as

$$E_Q = \sum_A E_A(Q_A) + \sum_{A>B} Q_A Q_B J_{AB}(R_{AB}) \tag{3.5-7}$$

where

    i. the energy of each atom (A) is considered to depend quadratically upon its charge Q_A

$$E_A(Q_A) = E_0 + \chi_A^0 Q_A + \frac{1}{2} J_{AA}^0 Q_A^2 \tag{3.5-8}$$

where

$$\chi^0 = (IP_A + EA_A)/2$$

and

$$J_{AA}^0 = (IP_A - EA_A) \tag{3.5-9}$$

are determined by the atomic ionization potential (IP) and electron affinity (EA) of the atom and

ii. The function $J_{AB}(R_{AB})$, determining the interaction between two atoms, is taken as the shielded Coulomb interaction between the two atoms (with radius, $R_A^0 + R_B^0$, determined from the crystal structures of the elements). As indicated in Figure 6, $J_{AB}(R_{AB})$ goes to the Coulomb form $(1/R_{AB})$ for large R, but to a finite limit as $R \to 0$.

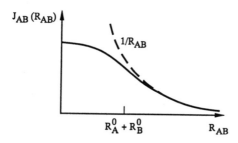

Figure 6. Form of the shielded potential, $J_{AB}(R_{AB})$.

Requiring that the charge distribution be in equilibrium (constant chemical potential) leads then to a set of linear equations

$$\sum_j A_{ij} Q_j = B_i \qquad (3.5 - 10)$$

for the unique equilibrium charges Q at a given geometry.

The QEq method is formulated so that there are *no free parameters*. The only data are the experimental ionization potential and electron affinity (which determine χ_A and J_{AA}) and the atomic radii [which determine the shielding of $J_{AB}(R)$ for small R]. The resulting charges are in good agreement with the PDQ charges from good quality wavefunctions.[31] This represents, we believe, a significant advance in simulations.

The IP_A, EA_A, and R_A data needed in QEq have been obtained directly from experimental atomic data and interpolated where necessary to define the parameters for every element of the periodic table through Lr (element 103).[33] We anticipate future generations of QEq where the parameters IP_A, EA_A, and R_A^0 are adjusted to fit the results of accurate *ab initio* calculations on a data base of molecules. This would allow fast prediction of *ab initio* quality charges.

The Charge Equilibration method (QEq) for finite molecules was developed jointly at Caltech and Molecular Simulations Inc. (Burlington, Massachusetts) and is commercialized in the BIOGRAF/POLYGRAF software for any atom up through Lr (element 103). With current programs, molecules with 1000 atoms are quite practical and there is no hard limit.

At the MSC the QEq method has been extended[32] to infinite systems (PBC), QEqX. Because of the slow convergence of Coulomb sums for infinite systems, we use Ewald approaches[15] with part of the sum in real space and part in Fourier (or reciprocal) space.

Since the chemical potential (6) depends on geometry, we can include the dynamic charge fluctuations in vibrational modes.[34] These dynamic charges are important for the intensities of IR and Raman modes and also affect the vibrational frequencies.

Some results[34] for charges in SiO_2 are given in Table 1.

Table 1. Charges for SiO_2 (reference 35).

Crystal	Pressure (atm)	Charge at Si	O-Si-O (deg)	Geometry SiO (Å)
Quartz	1	1.317	143.9	1.61
	21	1.331	138.8	1.63
	49	1.338	137.4	1.62
	61	1.346	134.2	1.60
Shistovite[a]	1	1.40		

[a]Six coordinate.

3.5.2 Valence Force Fields

Valence force fields represent the covalent chemical bonding part of the molecular interactions and involves terms depending upon bond, distances, bond angles, torsion angles, etc. We view such force fields in terms of a hierarchy ranging from very general but less quantitative at one end to very specific but accurate at the other.

3.5.2a Generic Force Fields

The standard force fields for organic and biological systems (MM2, MM3, AMBER, CHARMM) have evolved over many years of development. Yet these force fields deal primarily with just four atoms (H, C, N, O). To describe inorganic and organometallic systems, we must deal with the other 100 atoms of the periodic table. Our strategy is to use generic approaches, where the number of independent parameters are few and vary smoothly as a function of the rows and columns in the periodic table.

The first such force field[36] was DREIDING (developed in a collaboration between Caltech and Molecular Simulations, Inc.) which handles the 25 elements in the B, C, N, O, and F columns and the C, Si, Ge, Sn, and Pb rows of the periodic table (plus H and a few other elements). Here all parameters are independent of the particular bonds or molecules under consideration. Thus to do any molecule or crystal involving any combination of these elements, there are just two geometrical parameters per atom R^0 and θ^0, where R^0 is the atomic radius and θ^0 is the equilibrium angle of the hydride (e.g. CH_4, NH_3, SH_2, etc.). Similarly, the force constants are determined by very simple rules (e.g., K = 700, 1400, 2100 kcal/Å^2 for single, double, and triple bonds) that are independent of the atom. Despite the incredible simplicity, this force field gives quite accurate geometries for a wide variety of systems. Thus, using the first 76 accurate structures from the Cambridge database (including such groups as $-NO_2^-$, $-(SO_2)-$, $-PO_3^-$, $-(PO_2^-)$) the RMS error in atom positions is 0.235Å, with average errors of 0.009Å in bonds, 0.57 degrees in bond angles, and 0.22 degrees in torsions.[36]

Professor A. K. Rappè (Colorado State University), Dr. Mason Skiff (Shell Development Corp.), and the MSC have collaborated to develop a new generic force field suitable for any inorganic, organometallic, or organic molecule (any element of the periodic table). The first generation is called the Universal Force Field (UFF).[37] The only parameters in UFF are again atomic parameters. For the valence force field there are three parameters per atom rather than two! These parameters are R_i^0, θ_i^0 (as in Dreiding) plus an *atomic force constant* Z_i^0 defined such that the force constant for bond stretch is (11)

$$K_{ij} = 664 \frac{Z_i^0 Z_j^0}{(R_{ij}^0)^3} \qquad (3.5-11)$$

where the 664 converts units (energy in kcal/mol, distance in A) and where the bond distance

$$R_{ij}^0 = R_i^0 + R_j^0 - \delta R_{ij}(\chi) - \delta R_{ij}(bo) \qquad (3.5-12)$$

includes a correction for bond order,$\delta R_{ij}(bo)$, and a correction for electronegativity differences,$\delta R_{ij}(\chi)$. The force constants for angle bend $i - j - k$ are given by a similar expression involving Z_i^0 and Z_k^0. These atomic force constants, Z_i^0, vary smoothly along rows and columns, allowing them to be interpolated for elements with little data. The two vdw parameters are based on scaling of atomic quantities (HF polarizability and experimental IP) so that they are determined for every element. Thus with UFF one can predict structures for any combination of elements from H to Lr (element 103).

Combined with QEq (which can predict charges for any element from H through Lr), we now have a good start for simulating any combination of elements.

We expect to continue developing new generations of force fields using the ground rules *(i)* functional forms consistent with the physics and *(ii)* sufficient generality for applying to the complete periodic table. The current generation is good for geometries but barely adequate for vibrational frequencies. The new generations would focus on additional properties such as vibrational frequencies, polarization, etc.

3.5.2b Hessian Biased Force Fields[38,39]

At the other extreme from the generic force fields of the above section are complex force fields to accurately describe the properties of a specific class of molecules or polymers. Starting with the experimental geometries and vibrational frequencies for a molecule, it is straightforward to find force field parameters that provide a nearly exact fit to these data. However we do not consider this as adequate. The reason is that the experimental data do *not* define the vibrational properties of the molecule. To do this requires a full Hessian (second derivative matrix) of the molecule. Thus if

$$B_{i\alpha,j\beta} = \frac{1}{\sqrt{M_i M_j}} \frac{\partial^2 E}{\partial R_{i\alpha} \partial R_{j\beta}} \qquad (3.5-13)$$

is the mass weighted Hessian (atoms i, j and components α, β), then the vibrational states are given by

$$\mathbf{BU} = \mathbf{U}\lambda \qquad (3.5-14)$$

where $\mathbf{U}$ is a $3N \times 3N$ matrix, each column of which is the vibrational wavefunction for a mode, and $\lambda_{ij} = \nu_i^2 \delta_{ij}$, where ν_i is the vibrational frequency. It takes $\frac{1}{2}(3N)^2$ pieces of data to define $\mathbf{B}$ whereas the experimental frequencies provide only $3N$ (ignoring corrections due to rotation and translation), too few to determine the force field.

In order to determine B we must use modern quantum chemical programs (e.g. Gaussian 92) to calculate the full second derivative matrix (Hessian) directly from the electronic wavefunction. This provides the data to ensure that both the character (vibrational wavefunction) and frequency of each mode are described correctly. By fitting the parameters of the FF to the full Hessian, we can determine the functional forms and cross terms required in the force field. The difficulty is that accurate vibrational frequencies requires very accurate electronic wavefunctions. Thus with good basis sets the HF calculations are generally too expensive for large molecules (more than 20 atoms) and HF gives errors in the frequencies of 10-20%, which is far too large. The errors are smaller (approximately 5%) with such wavefunctions as MP2 but these become impractical for more than about 10 atoms. More complete wavefunctions that would yield accurate frequencies are generally not yet in a form that allows the analytic second derivatives needed by (13). Our current solution to this problem is the Hessian-Biased Force Field (HBFF)[38,39] which combines normal mode eigenstate information from HF theory with eigenvalue information from experiment. The procedure is to calculate B^{HF} from the HF wavefunctions and from this to obtain the vibrational eigenstates

$$B^{HF}U^{HF} = U^{HF}\lambda^{HF}.$$

We then construct the biased Hessian

$$B^{BH} = U^{HF}\lambda^{EXP}\tilde{U}^{HF} \qquad (3.5-15)$$

by combining U^{HF} from theory and $\lambda_i^{EXP} = \nu_i^{EXP}$ from experiment. B^{BH} has the property that

$$B^{BH}U^{HF} = U^{HF}\lambda^{EXP} \qquad (3.5-16)$$

That is, the eigenvalues are the experimental frequencies and the wavefunctions are the HF modes. Of course U^{HF} is not the exact description of the vibrational wavefunctions, but it is usually the best information we have.

We have used this HBFF approach to develop accurate force fields for many industrially interesting polymers, ceramics, semiconductors, and metals. Some of these applications are summarized in Sections 5 and 6.

3.5.3 Pseudoelectrons in Force Fields[42−44]

Standard force fields use springs to average over the electrons of quantum mechanics in describing structures and vibrationals of molecules. However there are many systems where the instantaneous response of the electron is essential in describing the properties. Rather than return to quantum chemistry to describe polarization effects (which would be to expensive for most simulations), we have found it possible to use pseudoelectrons in the force field to properly describe the polarization for polymers, metals, ceramics, and organometallics. We believe that force fields suitable for accurate prediction of the temperature behavior of moduli and other mechanical, dielectric, and optical properties will require use of such pseudoelectrons.

3.5.3a Covalent Shell Model[44]

In studying the piezoelectric and dielectric properties of poly(vinylidene fluoride) (PVDF), we developed the covalent shell model (CSM)[44] in which each atom is described with two particles [see (17)]. One (the nucleus, say C) possesses the mass and is connected to the valence springs of the standard force field theory. The other (the electron, e_C) is light (zero mass) and attached only to its nucleus with a spring constant (K_{Ce}) related to the charge (Q_{eC}) and polarizability (α_C).

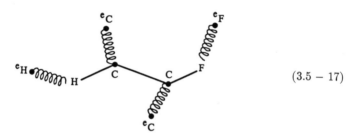

$$(3.5 - 17)$$

where

$$E_{Ce} = \frac{1}{2} K_{Ce} R_{Ce}^2 \qquad (3.5 - 18a)$$

$$K_{Ce} = \frac{Q_{eC}^2}{\alpha_c} \qquad (3.5 - 18b)$$

and

$$R_{Ce} = |R_C - R_e|$$

These atomic polarizabilities are obtained by fitting to the polarizability tensor from HF calculations on model systems. For example, with PVDF we used

$$(3.5 - 19)$$

Using CSM we can calculate the polarizability tensor of molecules, the dielectric tensor of crystals, and the piezoelectric constant tensor of crystals (see Section 5.4.2).

3.5.3b The Interstitial Electron Model for Metals and Alloys[42]

Most previous simulations on metallic systems have used two-body force fields, and the results have been quite poor. Recent advances based on GVB quantum chemical calculations[45] led to the Interstitial Electron Model (IEM) for force fields of metals in which the electrons are treated as classical particles, on the same footing as the ions.[42] This completely new type of force field for metals should allow accurate simulations for metals.

Interestingly, if instead of zero mass the interstitial electron is assigned the mass of free electron, the resulting vibrational frequencies for metals are in the range of 5–10 eV. These vibrations describe collective motions of the electrons and the energies are in the observed range for plasmons (which describe collective motions of the electrons) of metals. This suggests that the vibrational states of pseudoelectrons might have use beyond describing nuclear motion.

3.5.3c Force Fields for Organometallics[43]

Many organometallic systems involve molecular ligands such as cyclopentadienyl (Cp), allyl, ethylene, etc. To describe the structures and vibrational frequencies involving Cp, we find it necessary to define a pseudoatom at the center of the ring (denoted Cp),

$$\text{(figure)} \tag{3.5 - 20}$$

so that the FF includes terms such as Ti-Cp and Ti-Cp-C but no direct Ti-C terms. This is necessary in order to properly describe the stiff Ti-Cp vibration simultaneously with the loose bending and sliding modes of Cp with respect to Ti. We have defined this class of pseudoparticles as *center-of-mass* (CM) atoms and treat them as having no mass (they always adjust instantaneously to the current CM).[43]

3.6 Generalized London Force Fields for Simulations of Chemical Reactions[48]

Even the simplest chemical reaction say

$$\cdot H + D_2 \rightarrow HD + \cdot D \tag{3.6 - 1}$$

cannot be described with force fields of the form discussed above. The reason is that initially (reactants) there is a strongly attractive bonding interaction between the two D's whereas in the product there is a repulsive interaction between them. Similarly the H has a repulsive interaction with both D's at the beginning but has attractive interactions with one D in the product. Thus the pairwise interactions must change from repulsive to attractive and vice versa during the reaction. For a system such as (1) this leads to a barrier for the reaction with a height ($\sim$9.6 kcal/mol) that is about 10% of the bond energy ($\sim$104 kcal/mol). Thus there are clearly subtle interplays of bonding and antibonding factors where the bond is never really broken. This barrier is a direct consequence of the Pauli principle which allows only two electrons per orbital. As the third electron (of the H) approaches the D_2, it must remain orthogonal to the bond pair on the D_2 (to satisfy Pauli), leading to antibonding character and hence a barrier.

In contrast the FF for a MD study would use a normal bonding interaction (say Morse) between each particle and would incorrectly predict that the H_3 molecule is stable with a triangular geometry and a bond energy of $\sim$100 kcal/mol rather than a barrier of 10 kcal/mol. For proper description of the reaction surface, we *must* include the effects of the Pauli principle. Of course quantum chemistry (QC) calculations do this automatically. However

   i. the QC requires a great deal of effort for each geometry and we would like a way to fit the QC results with a general potential function

   ii. the QC is impractical for many cases of interest and we would like to have a way of predicting the interaction energies in the absence of QC.

We have developed a general procedure, the Generalized London Force Field (GLFF), for accomplishing both objectives.[48] This methodology is just being completed and has only been applied to three reactions; however, we expect to build this into a general procedure for all reactions and to develop routines for using this procedure with commercial simulation codes such as BIOGRAF/POLYGRAF.

The GLFF methodology is straightforward for any simple (three-electron) radical reaction, e.g.

$$\cdot A + X - B \rightarrow A - X + \cdot B. \qquad (3.6-2)$$

including surface reactions such as

$$(3.6-3)$$

The same formalism can also be used for simple four electron reactions, such as metathesis

$$
\begin{array}{ccc}
A-B & & A \quad\;\; B \\
& \Rightarrow & |\quad\;\; | \\
C-D & & C \quad\; D
\end{array}
\qquad (3.6-4)
$$

and insertion

$$
\begin{array}{ccc}
A-B & & A \quad\;\; B \\
& \Rightarrow & |\quad\;\; | \\
C=D & & C-D
\end{array}
\qquad (3.6-5)
$$

(e.g. Ziegler-Natta polymerization).

3.6.1 Spin Coupling

Consider the bonding in the reactant (Figure 7a), transition state (Figure 7b), and product as given in Figure 7:

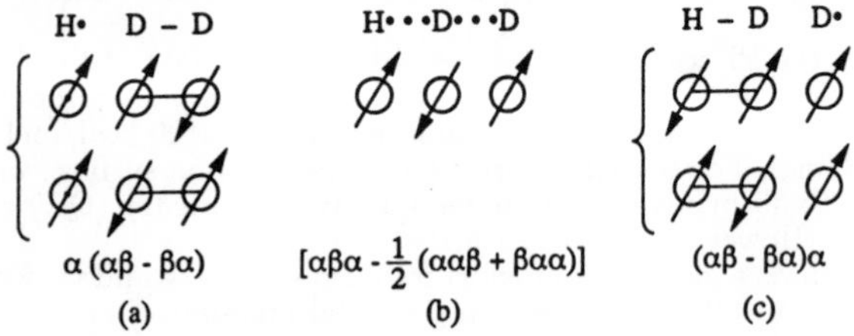

Figure 7. Illustration of changes in bonding and spin coupling. (a) Reactants, (b) saddle point, and (c) products.

The consequences of the Pauli principle can be directly expressed in terms of the spin couplings shown schematically in the middle row of Figure 7 and more explicitly in the bottom row.

For the reactant configuration the wavefunction is $\mathcal{A}\{\phi_l\phi_c\phi_r[\alpha(\alpha\beta-\beta\alpha)]\}$ where the electrons on the D_2 always have the opposite spins, leading to the singlet or bonding state of D_2. (Here ϕ_l, ϕ_c, ϕ_r refer to atomic orbitals on the left, center, and right atoms.) However the spin on the H is sometimes the same and sometimes different than the spin on the D atoms, leading to an $H-D$ interaction that contains both antibonding (or triplet) interactions and bonding (or singlet) interactions. Analysis of the QC wavefunction shows that the HD interaction is 75% triplet plust 25% singlet. The result of these triplet or antibonding terms is that the energy increases as H approaches D_2.

For the transition state the wavefunction is

$$\mathcal{A}\{\phi_l\phi_c\phi_r\left[\alpha\left(\alpha\beta-\beta\alpha\right)-(\alpha\beta-\beta\alpha)\alpha\right]\} \qquad (3.6-6)$$

which describes the resonance of $H-D\ D\cdot$ with $H\cdot\ D-D$. Analysis of the wavefunction shows that the outer orbitals are always triplet whereas the middle atom has interactions with the outer two atoms that are 75% singlet and 25% triplet.

Thus during the reaction the interactions between the H and central D changes for 25% singlet (reactant), to 75% singlet (transition state), to 100% singlet (product) and the other interactions charge correspondingly. Thus to include the effect of the Pauli principle, the simplest description is

$$E(R_1, R_2, R_3) = \sum_{i>j} E_{ij}(R_{ij}) \qquad (3.6-7)$$

where

$$E_{ij}(R_{ij}) = f_{ij}^S E_{ij}^S(R_{ij}) + f_{ij}^T E_{ij}^T(R_{ij}) \qquad (3.6-8)$$

Here f^S and f^T are the fractions of singlet and triplet character and E^S and E^T describe the bonding and antibonding two-body interactions. Rewriting

$$E^{cl} = \frac{1}{2}\left(E^T + E^S\right) \qquad (3.6-9)$$

$$E^x = \frac{1}{2}\left(E^T - E^S\right)$$

and

$$f_{ij}^S = (cos\frac{1}{2}\gamma_{ij})^2$$
$$f_{ij}^T = (sin\frac{1}{2}\gamma_{ij})^2 \qquad (3.6-10)$$

the energy (7) can be rewritten as

$$E = \sum_{i>j}\left[E_{ij}^{cl}(R_{ij}) + cos\gamma_{ij}E_{ij}^x\right]. \qquad (3.6-11)$$

The spin coupling angles γ_{ij} are related to each other (by $\pm\frac{2\pi}{3}$) so that there is only one degree of freedom. Requiring that γ be optimum, $\partial E/\partial\gamma = 0$, leads to

$$tan\gamma_{13} = \frac{\frac{\sqrt{3}}{2}(E_{23}^x - E_{12}^x)}{E_{12}^x + E_{23}^x - 2E_{13}^x} \qquad (3.6-12)$$

Since (12) depends only on $E^x(R_{ij})$, the optimum spin coupling is uniquely determined by the geometry. Thus the full effect of the Pauli principle is included by using (12) with (11), leading to

$$E_L = E_1^{cl} + E_2^{cl} + E_3^{cl} - \left[(E_1^x)^2 + (E_2^x)^2 + (E_3^x)^2 - E_1^x E_2^x - E_2^x E_3^x - E_2^x E_3^x - E_1^x E_3^x\right]^{\frac{1}{2}}.$$
$$(3.6-13)$$

which corresponds to the London equation.

3.6.2 Calculations

Obtaining E^{cl} and E^x in (11) from spline fits to very accurate quantum chemical calculations for H_2, leads to the potential energy surface potential energy surface[48] shown in Figure 8. This represents most of the important features of the exact energy surface, but the actual barrier (9.6 - 9.8 kcal/mol) is 20% smaller and the exact asymmetric stretch curvature is about half of the London value (see Figure 9).

The major errors in the London potential are due to (i) higher order overlap effects when all three atoms are close and (ii) including too much dispersion (vdw attraction) due to summing the pairwise dispersions. The Generalized London Force Field (GLFF) corrects for these effects by using correction terms based on the full VB wavefunction of the three-body system.

By carefully substituting the two-body VB energy expressions with $S_{ab}^2 \neq 0$ into the three-body VB energy expressions, also with $S_{ab}^2 \neq 0$, and retaining terms in overlap squared only, the first order overlap correction is found to be[48]

$$\Delta_3 = -\frac{1}{2}\left[(S_{13}^2 + S_{23}^2)E_{12}^x + (S_{12}^2 + S_{23}^2)E_{13}^x + (S_{12}^2 + S_{13}^2)E_{23}^x\right]. \qquad (3.6-14)$$

A good approximation to overlap is found by $S_i^2 = \delta^o E_i^x$, where δ^o is a constant. Using just this one parameter, the overestimate of the energy barrier is easily removed. The result is that the three-body corrections depend only on E^x.

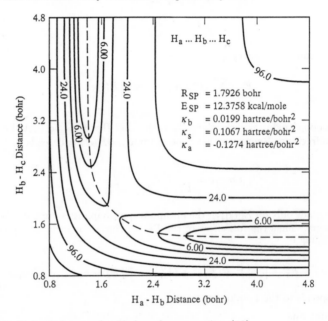

Figure 8. Potential curve for H_3 based on equation (13).

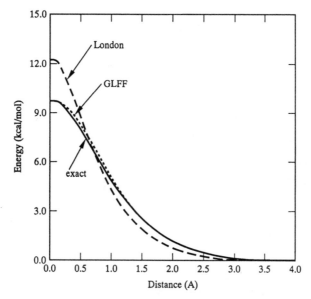

Figure 9. Energy for H_3 along the minimum energy path. L is the London potential using exact H_2 energies, E is the energy from the best QC calculations, and the solid line is the Generalized London force field. The origin is the saddle point.

The sum of exact two-body terms assumed in (7) and (8) includes too much London dispersion (vdw attraction) when two atoms are close. The dynamic correction effects (responsible for this attraction) for atoms one and two interfere with the 1-3 and 2-3 dispersion, leading to less attraction. This dispersion correction has the form

$$\Delta_{disp} = -\delta^d \sum_{i>j} E_{ij}^d \left(S_{ik}^2 + S_{jk}^2 \right) \qquad (3.6-15)$$

where E_i^d is the dispersion energy of pair i, and δ^d is a constant. A good estimate of the dispersion energy is found from the difference between the exact energy (experiment or configuration interaction) and the VB or GVB energy. This correction (15) prevents the potential energy surface from falling too quickly into either two-body valley.

These two corrections lead to an exact description of the barrier height and transition state geometry and to a much improved energy along the reaction path, as shown in Figure 9.

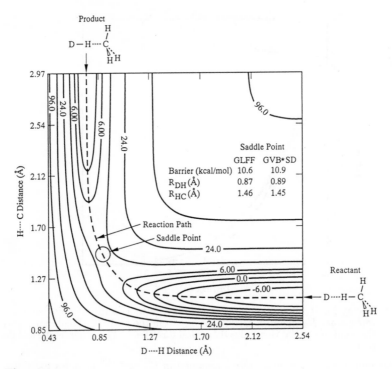

Figure 10.

The final FF has the form

$$E^{GLFF}(R_1, R_2, R_3) = \sum_{i>j} \left[E_{ij}^{Cl}(R_{ij}) + \cos\gamma_{ij} E_{ij}^x(R_{ij}) \right] + \Delta_3 + \Delta_{disp} \qquad (3.6-16)$$

This method has been applied to the

$$H \cdot \quad D - CD_3 \quad \rightarrow \quad H - D \quad \cdot CD_3 \qquad (3.6-17)$$

and

$$H_3C \cdot \quad D - CD_3 \rightarrow H_3CD \quad \cdot CD_3 \qquad (3.6-18)$$

reactions using $H \cdot \cdot H$, $H \cdots CH_3$, and $CH_3 \cdots CH_3$ two-body interactions from QC. The potential surface for (17) is shown in Figure 10. The barrier from GLFF is 10.6 kcal/mol, in excellent agreement with the 10.9 kcal/mol from extensive QC calculations on the reaction (17).

kcal/mol, in excellent agreement with the 10.9 kcal/mol from extensive QC calculations on the reaction (17).

3.6.3 Discussion

The approach outlined above can be used for any three electron (or four electron) exchange reaction,

$$A - B + \cdot C \rightarrow \cdot A + B - C$$

The first step is to obtain the singlet and triplet states for $A - B$ and for $B - C$. For best results we should do QC calculations at various distances for $A - B$ and $B - C$ separately. With no additional data on the $A - B - C$ system, the equation (13) predicts the full potential surface as in Figure 8. For H_3 this leads to a barrier $\sim 25\%$ too high and transition state bond distances about 3% too long. If there were no additional information, this could be used to simulate the reactions. Better yet we might do a quantum mechanical calculation at a single point near the predicted saddle point and another single point half way to dissociation. This would determine δ° and δ^d, allowing an accurate description of most geometries. Thus the GLFF allows a little bit of QC information to predict a great deal about the potential surface.

Alternatively with no QC data on the reacting units, we could use data, say, on the barrier height from experiment to estimate the overlap correction constant (δ°). In addition, if there were no QC data on the A-B bond and antibond, one could get a qualitative estimate by using a Morse curve for the singlet or bond state (which requires only R_e, k_e, and D_e) and an antimorse curve for the triplet or antibond state (no additional data).

3.7 Gibbs Canonical Molecular Dynamics[47]

Until recently the major use of molecular dynamics (MD) in simulations was as a means for searching conformational space to find stable structures or for examining the effects of thermal motions on the structure of a molecule. However future applications will focus more on the dynamics to obtain macroscopic materials properties as a function of temperature and pressure. Thus it has become important to formulate the dynamics so that accurate properties can be extracted.

Standard MD (Newton's equations) describes an adiabatic system (constant energy) whereas most experiments are carried out under conditions of constant external pressure and temperature. In order to use thermodynamic relationships to calculate the macroscopic properties for the dynamics, it is necessary that the phase space encountered during the trajectory generate a Gibbs ensemble (Boltzmann sampling) rather than the microcanonical ensemble for constant energy sampling. A breakthrough was made in the 1980's by Nosè,[49] and Parrinello and Rahman,[50] Nosè and Klein,[51] Andersen,[52] etc. who showed how to modify the Lagrangian or Hamiltonian in such a way that the new dynamical equations of motion lead to a canonical ensemble of momenta and coordinates (i.e. a Boltzmann distribution for constant pressure, constant temperature conditions).

We have implemented[47,53] these methods for Gibbs Canonical Dynamics [leading to Gibbs ensembles for TPN and TΣN (constant stress) conditions]. This allows the calculation of specific heat (C_p), thermal expansion tensor (α_i), and compliance tensor (S_{ij}) directly from thermodynamic response relations (volume-volume, volume-strain, strain-strain fluctuations) as a function of temperature and stress for crystalline, amorphous, and liquid systems.

For example, under constant temperature (T), constant stress (Σ) conditions, the specific heat, thermal expansion, and compressibility are given by

$$C_p = (\delta^2 H)/k_B T^2 \tag{3.7-1}$$

$$\alpha_p = (\delta V \delta H)/\bar{V} k_B T^2 \tag{3.7-2}$$

$$\kappa_T = (\delta^2 V)/\bar{V} k_B T \tag{3.7-3}$$

while the components of α and of the compliance (inverse elastic constants) are given by

$$\alpha_{ij,p} = (\delta e_{ij} \delta H)/k_B T^2 \tag{3.7-4}$$

$$S_{ijkl,T} = (\delta e_{ij} \delta e_{kl})/k_B T \tag{3.7-5}$$

where the δ^2 indicates the variance $[(\delta A \delta B) = < AB > - \bar{A}\bar{B}]$ over the trajectory. With periodic (in time) external stresses, this approach can be used to study modulus-temperature-time behavior (e.g., prediction of glass temperatures). For the time and size scales of interest this requires the use of massively parallel computers.

In addition to its use in calculating properties, the energy fluctuations inherent in canonical dynamics lead to very rapid equilibration of structures, as illustrated in Figure 11. Starting with the all trans polyethylene 40-mer, canonical dynamics leads to the expected 0.21:0.58:0.21 ratio of G^+, T, G^- in about 12 ps whereas with standard molecular dynamics the ratio has only relaxed to 0.21:0.67:0.12 in 920 ps![52] This occurs because in canonical dynamics the equations of motion result in energy flow in and out of the molecule just as in thermodynamic equilibrium, allowing the fluctuations in energy needed to overcome barriers. Consequently CMD may be very useful for protein folding and amorphous polymer applications.

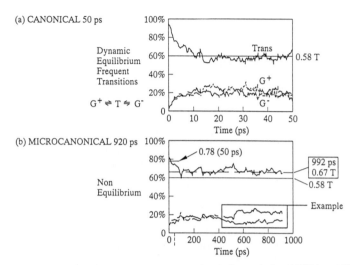

Figure 11. Dynamics for an isolated polyethylene chain, $(CH_3) - (CH_2)_{38} - (CH_3)$, including all 122 atoms. (a) 50 ps canonical dynamics (300K); this shows equilibrium to be attained in ~12 ps. (b) 920 ps of microcanonical dynamics (300K); this shows that the equilibrium distribution is not achieved after 920 ps.

3.8 Internal Coordinate Dynamics[58]

MD studies have generally used the Newtonian model in which all atoms in the molecule are treated as free particles and their accelerations are computed independently using Newton's law, i.e., $\ddot{x}_i = (1/M_i)F_i$. Hard constraints on bonds and angles are often used to eliminate certain degrees of freedom (e.g., CH vibrations, rotations about double bonds) and which have insignificant effect on long time-scale processes such as conformational changes in macromolecules (e.g., one might constrain all bonds and angles, allowing only torsions in the dynamics). It is particularly important to eliminate high-frequency degrees of freedom since they force the use of small integration step sizes, severely increasing the total cost for MD simulations.

In studying polymers and composites, it is essential to be able to easily convert the dynamics from fine level (all Cartesian coordinates), to intermediate level (all torsional degrees of freedom of the polymer, Rigid Unit Model of silicates), to coarse level (e.g., a rigid unit to describe the crystalline regions of a partially amorphous polymer plus torsional degrees of freedom for flexible interconnections), and then back. This hierarchy of rigidities internal coordinate dynamics allows the largest time scales and permits the simulation to focus on the relevant elements of the problem. To use such hierarchical rigidities for million atoms systems, it is essential to do the dynamics in the internal coordinates (e.g., torsions). In the Newtonian Model, doing the dynam-

constraints are generally treated with the iterative SHAKE algorithm in which the "unconstrained" equations of motion is used to compute an initial estimate of the change in the conformational state of the system.

3.8.1 Newton-Euler Inverse Mass Operator (NEIMO) Method[58-60]

The standard Newtonian methods for doing internal coordinate dynamics are completely inadequate for one million atom systems. The reason is that the equation of motion is

$$M\ddot{\phi} = T + C \qquad (3.8 - 1)$$

where ϕ is the vector of torsion angles, T is the vector of torques, C includes all long range interactions (expressed in cartesian coordinates) plus Coriolis terms, and M is the moment of inertia matrix. For a million atom polymer there might be 100,000 torsions so that M is a 100,000 by 100,000 matrix. To calculate the ϕ needed for dynamics we must solve

$$\ddot{\phi} = M^{-1}T + M^{-1}C, \qquad (3.8 - 2)$$

which involves inverting the M matrix, a process that scales as N^2 to N^3 depending on the methodology.

Jain and Rodriguez of JPL have developed the Newton-Euler Inverse Mass Operator NEIMO method,[59] which constructs the inverse directly in an order N process! This utilizes the Newton Euler formulation in which the velocities and forces of each unit are described relative to those of the connected units. Within this formalism it was possible to construct operator representations of M, ϕ, T, and C such that M could be *analytically* inverted. This leads to an expression for M^{-1} that can be evaluated in an order N process.

For the SGI computer we have adapted NEIMO to our CMM and MD programs to calculate dynamics of amorphous polymers.[60] We have recently implemented this methodology into MD programs for infinite crystals.[58]

4. Implementation on Scalable Massively Parallel Computers

Critical to atomistic simulation of the structure and properties of real materials is the development of superefficient software designed for scalable massively parallel computers. As a result we have focussed on methods that lead to natural localization of the simulation onto independent processors. We have developed a two step strategy. First we modify the theoretical methods so as to parallelize the *computational* part of the calculation but with little attention to memory allocation (no message passing). This is carried out using the 64 node KSR computer in the MSC (funded by NSF-GCAG). The KSR-64 can be used to validate the methods with production computations on 1 million atoms. This allows rapid implementation for testing of new methods and algorithms. After the methods are tested and optimized, we then migrate to the 512 node Intel Touchstone Delta at Caltech. This requires optimization of memory allocation and message passing. The strategy of decoupling the efforts of parallelization into two steps of parallelizing the computations and then parallelizing memory/message passing is, we believe, the optimum for developing efficient highly parallel computer software. The developments for the Intel are in collaboration with Professor Steve Taylor (Computer Science) at Caltech.

4.1 Quantum Mechanics

The pseudospectral method formally scales as N^3 and with cutoffs rapidly approaches N^2. This is the same scaling as for density functional methods (despite the presence of nonlocal exchange operators in PS-GVB). The current version of the PS-GVB (up to 2000 basis functions) is significantly faster than conventional *ab initio* codes (e.g., a factor of 3 for porphine as compared to GAUSSIAN 92) and comparable in performance to LDF codes such as DGAUSS.[16]

For electron correlation, we are exploring a new highly accurate algorithm[61] scaling in the N^3 range. This is a major breakthrough in the application of *ab initio* methods for electron correlation to large systems, rendering them competitve in computational costs with LDF methods while allowing accuracy to be improved systematically. The methods employ GVB techniques wedded to a localized version of perturbation theory. The PS implementation provides orders of magnitude speedup of both the GVB and perturbation theory algorithms.

A key feature of the methodology from the standpoint of parallelization is that a numerical grid is employed as well as a standard Gaussian basis set. This ensures that, even in a SIMD framework, a large set of uniform and readily distributed work is always available. This can be contrasted with conventional *ab initio* electronic structure technology, where the indexing of the two electron integrals becomes very complex, leading to significant problems in parallelization.

The parallelization of the Hartree-Fock part of the algorithm (PS and LDF) is most straightforwardly accomplished by placing a subset of grid points and each processor. The contribution to the Fock matrix from each section of the grid is essentially independent and hence no interprocessor communication is required until one is ready to assemble the final Fock matrix. There are load balancing issues involved because of cutoffs on the grid, which could degrade performance by as much as a factor of 2. Aside from this, the overall ratio of communication to computation is such that latency times and bandwidth should not be major issues.

We are initially implementing a purely grid distributive approach and will address the load balancing problems as necessary. There are also some parts of the algorithm which are not grid based. The most serious issue here is diagonalization of the Fock matrix. We have developed a new large system Lanczo eigensolver for all occupied orbitals which is about 4 times faster than standard methods and should be efficient for massively parallel platforms. An interesting possibility is the suggestion by Baroni and Giannozzi[18] of iterative methods that avoid diagonalization, leading to order N as the slowest step.

The algorithm for orbital optimization in GVB calculations is very similar to Hartree-Fock. For GVB-PP (perfect pairing), there are several Coulomb and exchange operators to assemble rather than one. This necessitates more memory per node (the demands are not excessive for modern machines) but otherwise breaks up into localized pieces just the same way as Hartree-Fock.

4.2 Molecular Dynamics

The CMM method leads to a natural localization in which a physical part of the system is associated with each node. Thus with 512 nodes ($8^3 = 512$) each third level cell is on a physical node and deeper levels are quite local. The most straightforward parallel implementation of RCMM is to configure the multiprocessors as a 3-dimensional (3D) array. Thus we partition the 3D physical space into similar equally-sized rectangular boxes and each box is mapped into one processor in the processor array. This box is further divided into smaller boxes (cells) until each cell contains about 4 atoms, so that the interactions generated by the atoms in this processor is done in the same straightforward way as for the sequential computer. Most of the long range interactions from other boxes (of atoms) are contained in the Taylor series expansion. This requires 10 or 20 floating point numbers (quadrupole or octupole) which are easily passed among the processors. The most difficult part are the interactions in nearest neighbor cells to the cell we are doing, the intermediate cells (they are near the box boundaries). Physically these intermediate cells contain a small part of the space, but they are non-negligible because they are relatively close.

The NEIMO procedure is local, allowing efficient parallelization. However the localization (assignments of atoms to processors) optimum for CMM is different than the localization optimum for NEIMO (assignment of chains to processors). These conflicting requirements are balanced by considering the chains within each processor as separate and correcting for the boundary conditions during the NEIMO iteration.

5. Application to Mechanical and Structural Properties of Polymers, Blends, and Composites

Developments in polymers and composites are at the heart of many new industrial materials technologies. Recent advances include environmentally stable high-temperature thermoplastics (e.g., PEEK) for structural applications, photoresists (for fabrication of ultra-large scale integrated circuits), flexible electronic conductors, and highly polarizable materials (ferroelectrics, piezoelectrics, nonlinear optical materials). Even so, the design of new materials and the development of new processes for preparing them is seriously impeded by our current inadequate understanding of how atomic-level structure leads to macroscopic properties. Indeed because useful polymers are generally disordered, the current experimental methods generally provide only hazy guidelines concerning local structure. The design of such new industrial materials and the development of new processes for preparing them is seriously impeded by our current inadequate understanding of how atomic-level structure leads to the structures and macroscopic properties. To design new materials we need the ability to use atomic-level simulations to predict detailed atomic-level structures and properties and to describe the dynamical processes as the system is stressed, extruded, etc. Our goal here is to develop the tools required for such simulations and to apply these tools to industrially important materials critical problems.

As a first step toward simulations of polymers, we developed the ability to calculate structural and mechanical properties of polymers including elastic constants (e.g., Young's modulus, Poisson ratio, compressability), thermodynamic properties (Cp, S, H, F, etc.), surface energies, phonon dispersion curves (vibrational states), and stress-strain curves for finite deformations.[1-3] In addition, we developed tools for such properties as piezoelectricity and hyperpolarizabilities (for nonlinear optical properties).

5.1 Moduli

Particularly important for most applications are the elastic constants that determine the basic mechanical properties (e.g. Young's Moduli). To illustrate the accuracy from atomistic calculations[62] we summarize in Table 3 the best experiments on PE for the Young's modulus

$$E_c = \frac{1}{S_{cc}} = \frac{\Delta \sigma_c}{\Delta c_c},$$

along the chain direction. The results range from 213 to 358 GPa (at room temperature)! The theory leads to 318 GPa which we estimate to be low by ~4 GPa, leading to an "exact" answer of 322 ± 9 GPa. This is well within experimental error from inelastic neutron scattering (329 ± 15) and one can understand that the X-ray values might be off (uncertainties in the local internal stresses).

However one might wonder about why the Raman results (358 ± 25 and 290 ± 5) do not agree with each other or with neutron scattering. The two Raman values are based on measurements of the accordian band of linear alkanes ($C_n H_{2n+2}$) from $n = 10 - 90$, extrapolation to $n = \infty$ (see Table 4), and then correcting for chain-chain interactions. Our theory[62] agrees with the observed Raman modes for $n = 10 - 16$, leading to frequencies too low by 0.7% (see Table 4). Our calculations also lead directly to a Young's modulus 318 GPa for the crystal. Based on the comparison with the Raman data we expect this value to be low by 1.4%, leading to an estimated $E_c = 322 \pm 9$ GPa. Thus the disagreement between theory and the Raman experiments is due to the extrapolation of the Raman data to the crystal! Thus even on one of the most studied polymers, PE, the theory can now provide better data than experiment. For the other elastic constants, the experimental data are in far worse shape, and we believe that the theory provides the first reliable sets of elastic properties for crystalline polymers. The values for PE are given in Table 5 for[62] PE. As it becomes routine to evaluate the effect upon elastic constants due to changes in the packing, conformation, and composition of polymers, we should be able to design polymers with their properties tailored for specific applications. For example, it should be possible to develop a polymer with zero or negative Poisson ratios (expands perpendicular to the direction of tension).

A major activity at the MSC is in developing for additional classes of polymers the high quality force fields necessary to such design activities.

Table 3. Young's Moduli (in GPa) of polyethyene.[a] See reference 62 for explicit references.

	0 K	77K	300 K
(a) E_c (chain axis)			
theory (MCXX,n-butane)	337	332	318
experiment			
direct			
X-ray			235
X-ray			213-229
dynamic tester	$(324 \pm 30)^c$	288 ± 10	
spectroscopic			
neutron scattering			329
Raman			358 ± 25^d
Raman			$290 \pm 5^{b,d}$
(b) E_b ($\perp$ chain, long axis)			
theory (MCXX,n-butane	9.4	9.1	6.1
experiment			
X-ray			3.2
X-ray			2.5
X-ray			5.0
(c) E_b ($\perp$ chain, short ax)			
theory (MCXX, n-butane)	9.0	7.6	6.0
experiment			
X-ray			3.9
X-ray			1.9

[a]The theoretical values are calculated from $E_c = (S_{11})^{-1}$, $E_c = (S_{22})^{-1}$, $E_c = S_{33})^{-1}$, where $S = C^{-1}$ is the compliance matrix. [b]Extrapolation from isolated n-alkane chains corrected for interlamellar interactions. [c]Estimated. [d]Extrapolation from isolated n-alkane chains.

Table 4. Comparison of theoretical and experimental frequencies (cm^{-1}) for the accordian (Raman) mode for n-alkanes (C_pH_{2n+2}). See reference 62 for explicit references.

n	Theory (MCXX)	Experiment least squares[a]	liquid	solid
4	424	473	429 (5)	425 (4)
5	402	411	400 (7)	406 (3)
6	372	358	370 (4)	373 (3)
7	305	313	310 (6)	311 (5)
8	278	278	279 (5)	283 (3)
9	243	249	248 (3)	249 (2)
10	225	225	230 (3)	231 (3)
11	198	206		
12	189	189	195 (1)	194 (2)
13	177	175		
14	163	163		
15	153	153		
16	143	144		140 (3)

[a] Based on the analytic least-squares fit to the observed frequencies.

Table 5. Elastic constants for PE. All quantities in GPa unless otherwise indicated (reference 62).

	4 K	77 K	213 K	303 K	411 K
C_{11}	14.9	13.3	10.6	8.3	3.9
C_{22}	12.9	11.2	9.6	8.2	5.4
C_{33}	338.2	333.2	325.7	318.4	306.5
C_{12}	7.8	6.9	5.4	4.3	3.7
C_{13}	2.2	1.8	1.2	0.7	0.09
C_{23}	4.2	3.5	3.0	2.5	1.9
C_{44}	4.8	4.0	3.5	3.0	2.5
C_{55}	2.9	2.5	2.1	1.7	1.4
C_{66}	6.7	6.1	4.7	3.6	2.3
E_a	10.2	9.1	7.5	6.1	1.3
E_b	8.8	7.6	6.8	6.0	1.8
E_c	336.8	332.1	324.7	317.6	304.6
$\beta_a(GPa^{-1})$	0.0945	0.107	0.130	0.161	0.258
$a_a(\text{Å})$	7.121	7.155	7.287	7.413	7.706
$b_a(\text{Å})$	4.851	4.899	4.918	4.942	4.936
$c_a(\text{Å})$	2.548	2.5473	2.5473	2.5473	2.5473
$\theta_a(\text{deg})$	42.4	42.5	42.3	42.2	40.4

5.2 Stress-Strain Curves[62]

It is now routine to optimize all internal geometric parameters and all lattice parameters in the presence of fixed external stresses.[62] In this way we have been able to calculate the stress-stress curves as a polymer is stretched to fracture as in Figure 12 or sheared as in Figure 13.

Beyond the ultimate stress of the polymer (0.2 and 0.45 GPa for the cases in Figure 14), the material fractures. In order to estimate the accuracy in such predictions, consider the lattice energy of crystalline polyethylene at 0K, that is the energy to pull polyethylene crystal apart into isolated polyethylene chains. The calculations lead to 1.8701 kcal/mol (per CH_2) which can be compared with experimental values of 1.838 $\pm$ 0.032 kcal/mol.[62] Thus we appear to be high by about 2% in the total chain-chain interactions.

The critical shear strength parallel to the chains from Figure 13 is 0.035 GPa. This should be similar to the shear strength of long oriented alkane fibers. Indeed the predicted critical stress compares well with the value of 0.028 GPa for linear alkanes deduced from high stress tribology experiments.

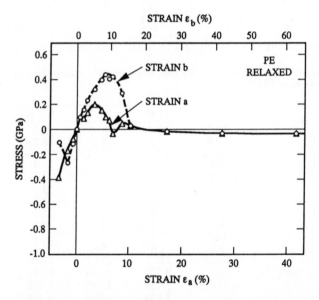

Figure 12. Stress versus strain for finite deformation in the a and b directions of polyethylene crystal (MCXX n-butane force field). Cell parameters perpendicular to the deformation direction are optimized.

Similar studies are in progress on a number of polymer systems including alkane polymers (polypropylene)[63], nylons (nylon 66)[64], PEEK[65], PVDF (polyvinylidene difluoride)[44], POM (polyoxymethylene)[66], phenol carbonates[66], and PTFE (polytetrafluorethylene)[67]. The ability to predict such critical stress should be most valuable in designing new polymers.

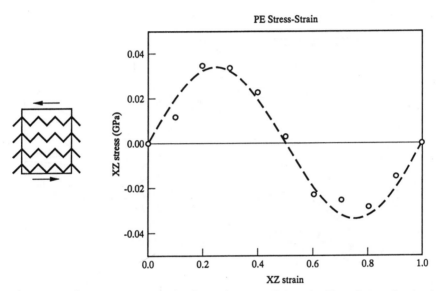

Figure 13. Stress versus strain for finite shear parallel to the fiber chains of polyethylene crystal. The calculated points are shown with circles, the dashed line shows a simple fit to a sine function.

5.3 Thermodynamic Properties

The thermodynamic properties of polymers, can be calculated directly from the vibrational states of phonons of the crystal. Some typical results[62] for PE are shown in Figure 14. Such phonon dispersion data will, we believe, prove useful in predicting dynamical properties of polymers. Indeed the maximum energy for the lattice modes in directions perpendicular to the chain axis seems to correlate with the glass temperature (T_g). We intend to obtain such polymer dispersion curves for a number of polymers to establish a data base useful for finding such correlations. Since experimental phonon dispersion curves are generally not available [IR and Raman leads to some data for states at $k = 0$ (Γ point); inelastic neutron scattering leads to some data for longitudinal acoustic modes], such connections have not been previously examined).

Given the phonon states we can predict the specific heat, entropy, and free energy versus temperature. Typical results[62] are shown in Figure 15 for PE. The prediction of such thermodynamic data for new polymers will, we believe, prove useful in designing activities.

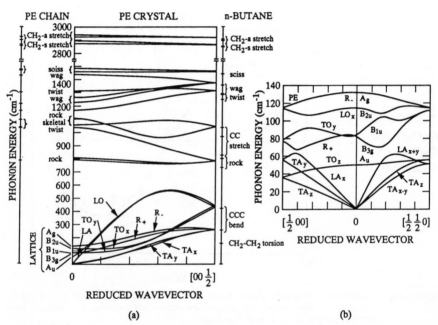

Figure 14. Phonon dispersion curves of polyethylene crystal (MCXX n-butane force field) (a) in the chain direction (c axis) and (b) in the [100] and [110] direction. In (a), frequencies of the isolated infinite chain at k = 0 and those of n-butane are also shown. (From reference 62.)

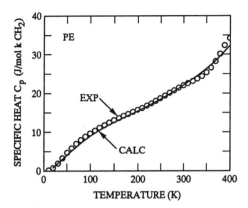

Figure 15. Calculated C_p and experimental C_p versus temperature. (From reference 62.)

5.4 Poly(vinylidene Fluoride)[44]

Poly(vinylidene fluoride) (PVDF) polymer (1) has piezoelectric and mechanical properties that make it technologically interesting.

$$\left(CH_2 \diagdown CF_2 \right)_n \tag{5.4-1}$$

Thus applying a voltage across a block of the material causes a strain and applying a stress induces a voltage. In addition PVDF exhibits four well studied crystal forms, with piezoelectric properties observed in two. Thus this material serves as a good test for molecular simulations since one must account for both the stability and nearly equal cohesive energies of several forms in addition to the mechanical and electrical properties.

In reference 44 we developed two forces fields (MSXX and MSXX/S) for PVDF based on a combination of first principles quantum chemical (QC) calculations and experimental phonon frequencies of form I. Both force fields include the cross terms required for accurate vibrational (phonon) frequencies. MSXX/S also includes the Covalent Shell Model (CSM) for describing polarizabilities. We do not readjust the FF parameters to fit observed crystalline properties (structure, elastic constants). Thus the accuracy in predicting known properties can be used to assess the accuracy for prediction of unknown structures and properties.

These force fields were used to study the stabilities and properties of four observed forms of PVDF plus five additional forms which we find to be stable (including the antipolar form of II suggested by Lovinger). The calculated properties include elastic constants (C_{ij}), piezoelectric moduli (d_{ij}), and dielectric constant tensor (ϵ_{ij}).

5.4.1 Structures, Energies and Processing

Table 6 summarizes various characteristics of the four observed structures of PVDF plus the five other stable structures that we studied. All nine of these structures are stable (positive definite elastic constant tensor and all phonon frequencies positive).

The three common notations for the four observed structures are I,II,III,IV or $\beta, \alpha, \delta, \gamma$ or I, II, III, II_p respectively. In order to simplify reference to all forms we will use I, II, III to indicate conformation (all T, TG, or $TTTG$, respectively), followed by superscripts p and a to indicate parallel or antiparallel orientation of the chains (about the chain axis), followed by u and d to indicate up-up or up-down. Thus I becomes I_p (or T_p); II becomes II_{ad} (or TG_{ad}); III becomes III_{pu} (or T_3G_{pu}); and IV becomes II_{pd} (or TG_{pd}). Form II normally has a statistical distribution of up-up and up-down, so that II becomes II_a while IV becomes II_p. All nine forms are shown in Figure 16.

For both force fields, MSXX and MSXX/S, we find that the differences in total energies of the four observed forms are quite small (within 1 kcal/mol per monomer unit) while the two forms never observed are calculated to be over 1 kcal/mol. This is encouraging because we would expect that experimentally observable forms should be with 1 kcal/mol. This suggests that the potentials are accurate.

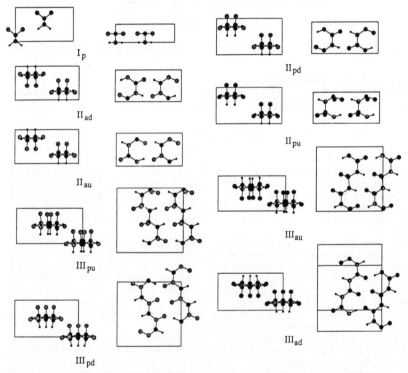

Figure 16. Crystal packing of the nine stable crystal forms for PVDF.

Table 6. Names and characteristics for various forms of crystalline poly(vinylidene fluoride), PVDF.

	I_p	I_a	II_{pu}	II_{pd}	II_{au}	II_{ad}	III_{pu}	III_{pd}	III_{au}	III_{ad}
Names:										
New Simple	I_p	I_a	II_{pu}	II_{pd}	II_{au}	II_{ad}	III_{pu}	III_{pd}	III_{au}	III_{ad}
New Complete	T_p	T_a	TG_{pu}	TG_{pd}	TG_{au}	TG_{ad}	T_3G_{pu}	T_3G_{pd}	T_3G_{au}	T_3G_{ad}
Roman	I	–	(IV)	IV	(II)	II	III		V	
Roman'	I	–	(I_p)	II_p	(II)	II	III_p		III_a	
Greek	β		(δ)	δ	(α)	α	γ			
Conformation	T	T	$TGT\overline{G}$	$TGT\overline{G}$	$TGT\overline{G}$	$TGT\overline{G}$	$T_3GT_3\overline{G}$	$T_3GT_3\overline{G}$	$T_3GT_3\overline{G}$	$T_3GT_3\overline{G}$
Alignment										
||chain	–	–	u-u	u-d	u-u	u-d	u-u	u-d	u-u	u-d
⊥ Chain	p	a	p	p	a	a	p	p	a	a
Polar										
||Chain	No	No	Yes	No	Yes	No	Yes	No	Yes	No
⊥ Chain	Yes	No	Yes	Yes	No	No	Yes	Yes	No	No
Space Group	C_{m2m}		C_c	$P2_1cn$	$Pca2_1$	$P2_1/c$	C_c	$P_{na}2_1$	$Pca2_1$	$P2_1/c$
	C_{2v}^{14}		C_s^4	C_{2v}^9	C_{2v}^5	C_{2h}^5	C_s^4	C_{2v}^9	C_{2v}^5	C_{2h}^5
Energy (MSXX)	0		0.68	0.69	0.85	0.78	0.30	1.90	0.46	1.20
(MSXX/S)	0			0.67		0.72	0.46		0.64	
Observed	Yes	No	Stat	Stat	Stat	Stat	Yes	No	No	No
Piezoelectric	Yes	No	Yes	Yes	Yes	No	Yes	Yes	Yes	No
$d_{\perp\perp}^a$ (MSXX/S)	-18.8			-2.9		0	-2.7		0	
Young Mod (E_c)										
MSXX	293			163		153	113		102	
MSXX/S	277			150		141	107		92	
Compress (β)										
MSXX/S	15.2			13.6		13.6	14.7		13.5	

ᵃ Piezoelectric modulus connecting a change of polarization in the direction perpendicular to the chain with a tension stress in that direction.

In Figure 17 we sketch the relationships between various forms of PVDF and processing conditions, including our calculated energies. We see here that the energetics are quite consistent with observed processing. Thus cooling from the melt leads to II_a (E = 0.72 kcal/mol) and drawing at room temperature leads to I_p (E = 0.0). Poling (application of external electric fields) of II_{ad} leads to the polar form II_{pd} (with E = 0.67). High temperature annealing from II_{pd} leads to III_{pu} (E = 0.46). Poling from II_{pd} (E = 0.67) or drawing from III_{pu} (E = 0.46) leads to I_p (E = 0).

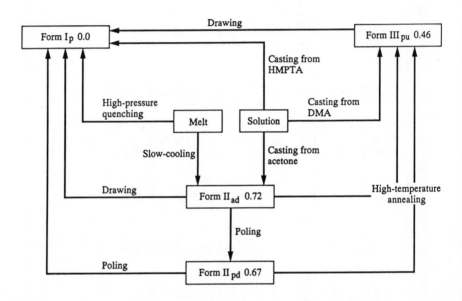

Figure 17. Processing relations between form of PVDF. The calculated energies are in kcal/mol CH_2CF_2 (see reference 44).

Table 7. Young's modulus in the chain direction (E_c) and bulk modulus (β) for all forms of PVDF in GPa. Calculations were carried out using the experimental cell parameters (exp) and the optimized cell parameters (opt) (reference 61).

Form	I_p		II_{ad}		III_{pu}		II_{pd}		III_{au}	
	exp	opt	exp	opt	exp	opt	exp	opt	exp	opt
MSXX										
E_c	282.5	292.7	151.2	152.9	107.7	113.4	156.0	162.8	-	101.7
β	9.5	14.8	11.7	12.4	11.7	13.3	9.8	12.1	-	12.5
MSXX/S										
E_c	265.2	277.2	136.8	140.9	97.2	107.2	139.3	150.2	-	92.3
β	8.8	15.2	10.5	13.6	10.8	14.7	8.6	13.6	-	13.5

Table 8. Elastic stiffness constants (GPa) of PVDF calculated at experimental structures (except for form III_{au}, where the optimized structures is used), (reference 61).

	MSXX/S					
Form I_p	23.7	2.4	4.7	0.0	0.0	0.0
	2.4	11.8	3.1	0.0	0.0	0.0
	4.7	3.1	266.8	0.0	0.0	0.0
	0.0	0.0	0.0	3.7	0.0	0.0
	0.0	0.0	0.0	0.0	5.2	0.0
	0.0	0.0	0.0	0.0	0.0	4.0
Form II_{ad}	22.4	6.3	−7.4	0.0	−3.1	0.0
	6.3	13.1	7.9	0.0	0.0	0.0
	−7.4	7.9	148.4	0.0	0.3	0.0
	0.0	0.0	0.0	3.8	0.0	0.2
	−3.1	0.0	0.3	0.0	7.6	0.0
	0.0	0.0	0.0	0.2	0.0	5.3
Form III_{pu}	19.2	6.3	−2.7	0.0	−0.6	0.0
	6.3	14.4	9.8	0.0	−0.6	0.0
	−2.7	9.8	106.9	0.0	−1.0	0.0
	0.0	0.0	0.0	2.7	0.0	−0.4
	−0.6	−0.6	−1.0	0.0	2.8	0.0
	0.0	0.0	0.0	−0.4	0.0	7.7
Form II_{pd}	22.7	4.6	−7.2	0.0	0.0	0.0
	4.6	10.2	5.5	0.0	0.0	0.0
	−7.2	5.5	146.8	0.0	0.0	0.0
	0.0	0.0	0.0	3.9	0.0	0.0
	0.0	0.0	0.0	0.0	7.6	0.0
	0.0	0.0	0.0	0.0	0.0	5.1
Form III_{au}	22.3	8.1	15.8	0.0	0.0	0.0
	8.1	19.0	−2.6	0.0	0.0	0.0
	15.9	−2.6	108.1	0.0	0.0	0.0
	0.0	0.0	0.0	4.1	0.0	0.0
	0.0	0.0	0.0	0.0	2.6	0.0
	0.0	0.0	0.0	0.0	0.0	7.5

5.4.2 Elastic, Dielectric and Piezoelectric Properties

Elastic, dielectric and piezoelectric properties were calculated for all forms using analytic second derivatives of energy at the optimized structures.

The elastic stiffness constants (C_{IJ}) and compliance constants (S_{IJ}) are defined as

$$\sigma_1 = C_{IJ}e_J \qquad (5.4-20)$$

$$e_J = S_{JI}\sigma_I \qquad (5.4-21)$$

where e_J are components of strain and σ_I are components of stress (repeated indices are summed, Einstein convention). Here I,J=1,2...6 denotes xx,yy,zz,yz,zx, and xy, respectively.

The bulk modulus (β) is defined by

$$\beta^{-1} = \sum_{I,J=1}^{3} S_{IJ}$$

and the Young's modulus in the chain direction is defined by

$$E_c = \frac{\sigma_c}{e_c} \qquad (5.4-22)$$

where e_c and σ_c are strain and stress in the chain direction. Tables 7 or 8 shows the results calculated for the five forms.

Based on E_c, the strength of these crystal is in the sequence $I_p > II_{pd} > II_{ad} > III_{pu} > III_{au}$.

The MSXX/S, which includes shell polarization effects leads to a decrease in the elastic constants by about 10%. These changes arise from changes in the valence parameters for MSXX/S.

The experimental Young's modulus of the form I_p crystal in the chain direction is 177 GPa. In this experiment, stress is applied along the chain direction and a shift of the characteristic X-ray diffraction spots associated with periodic structures in the polymer is observed. The elastic modulus is then calculated under the assumption of homogeneous stress. This value is much smaller than our calculated value of 283 GPa. We believe that the discrepancy is caused mainly by the difficulty in extracting the properties of a pure crystal from samples which have a mixture of the crystalline and amorphous regions. Similar results were found for polyethylene where the X-ray methods led to $E_c \sim 235$ GPa whereas neutron inelastic scattering, Raman, and theory all lead to 322 ± 9 GPa.

Including both strain and electric field as independent variables leads to the relations

$$\sigma_I = C_{IJ}e_J - g_{kI}\mathcal{E}_k \qquad (5.4-23a)$$

$$\mathcal{P}_k = g_{kJ}e_J + a_{kl}\mathcal{E}_l \qquad (5.4-23b)$$

where $\mathcal{E}$ is the external electric field and $\mathcal{P}$ is the dielectric polarization. Here **g** is the piezoelectric constant tensor and **a** is the dielectric susceptibility tensor at constant strain. Alternatively, considering both stress and electric field as independent variables leads to

$$e_I = S_{IJ}\sigma_J + d_{kl}\mathcal{E}_k \qquad (5.4-24a)$$

$$\mathcal{P}_k = d_{kJ}\sigma_J + b_{kl}\mathcal{E}_l \qquad (5.4-24b)$$

where **d** is the piezoelectric modulus tensor and **b** is the dielectric susceptibility tensor at constant stress. The **a** and **b** tensors are used to calculate the dielectric constant tensors at constant strain ($\epsilon^e = 1 + 4\pi\mathbf{a}$) and at constant stress ($\epsilon^\sigma = 1 + 4\pi\mathbf{b}$). The difference between the dielectric tensor at constant stress and that at constant strain is given by

$$\epsilon_{ij}^\sigma - \epsilon_{ij}^e = 4\pi \sum_K d_{iK}g_{jK} \qquad (5.4-25)$$

Table 9 shows ϵ^σ for all forms and both force fields. For form I, the axes are transformed such that the orientation is the same one used in the experiments (1 axis is parallel to chain direction c, 2 axis is parallel to a, and 3 axis is parallel to b and the polarization direction). As expected MSXX/S has a dramatic effect. The experimental determination of the properties for the crystalline form is quite difficult since the dielectric constant of the amorphous region is quite large and depends strongly on temperature and frequency. Measurements using the oriented films of form I crystal give $\epsilon_3 = 3.6$ ($-106°$C, 0.065 MHz), 3.7 ($-100°$C, 0.049 MHz), and 3.1 ($-102°$C, 0.059 MHz) which should be compared with the calculated value of $\epsilon_3^\sigma = 2.44$ for MSXX/S. The calculated value is smaller than experiment but this may be due to the amorphous regions in the experimental sample.

Table 9. Dielectric constants of PVDF at constant stress (ϵ^σ) calculated at experimental structure (except for form III_{au}, where the optimized structures is used). (Reference 61).

	MSXX/S		
Form I_p	2.8716	0.0000	0.0000
	0.0000	2.8599	0.0000
	0.0000	0.0000	2.4405
Form II_{ad}	2.2338	0.0000	−0.1434
	0.0000	2.1385	0.0000
	−0.1434	0.0000	2.1603
Form III_{pu}	2.4571	0.0000	−0.1860
	0.0000	2.1608	0.0000
	−0.1860	0.0000	2.4544
Form II_{pd}	2.3340	0.0000	0.0000
	0.0000	2.1501	0.0000
	0.0000	0.0000	2.2228
Form III_{au}	2.1545	0.0000	0.0000
	0.0000	2.3010	0.0000
	0.0000	0.0000	2.2507

Table 10 shows the piezoelectric moduli (d) calculated at the experimental cell parameters for forms. The experimental value of d_{33} of form I_p crystal is -20 ± 5

pC/N, which is quite close to the calculated value with the shell model of -18.8 pC/N. Also, with experimental value of d_{31} is reported to be much smaller than d_{33}, which is consistent with our results, $d_{31}/d_{33} = -0.5/18.8 = -0.02$.

Table 10. Piezoelectric modulus tensor d_{iJ} (pC/N) where J=1 to 6 and i=1,2,3. Calculations are at the experimental structures (except for form II_{au}), where the optimized structures is used). (Reference 61.)

	MSXX/S					
Form I_p	0.0	0.0	0.0	0.0	−41.9	0.0
	0.0	0.0	0.0	−16.8	0.0	0.0
	0.57	−1.3	−18.8^a	0.0	0.0	0.0
Form III_{pu}	4.4	4.2	−2.7	0.0	−1.2	0.0
	0.0	0.0	0.0	12.0	0.0	5.1
	3.3	−1.6	2.2	0.0	27.6	0.0
Form II_{pd}	0.0	0.0	0.0	0.0	−9.1	0.0
	0.0	0.0	0.0	−6.1	0.0	0.0
	1.8	−2.4	−2.9	0.0	0.0	0.0
Form III_{au}	0.0	0.0	0.0	0.0	1.15	0.0
	0.0	0.0	0.0	3.85	0.0	0.0
	3.26	−2.54	0.71	0.0	0.0	0.0

^{a}Experimental value is -20 ± 5.

5.5 Surfaces and Interfaces[67]

Critical to most applications of polymers are the properties at surfaces and interfaces. Thus, in a composite we need strong bonding between fiber and matrix in order to transfer shear stress: however, to obtain a tough (nonbrittle) material, we might want this interface to slip so as to absorb energy under sufficiently large stresses. On the other hand, for pressure-sensitive adhesives, the relevant property is tackiness, that is, the ability to stick under light pressure to various surfaces. The most important properties about surfaces and interfaces are the surface energy and the surface tension, which directly affect wettability and adhesion.

Unfortunately there is very little reliable data about how the atomic structure of a polymer affects the properties at surfaces and interfaces. The surface science techniques so useful for characterizing surfaces of semiconductors and metals have been of limited usefulness for polymers. [These techniques all involve electrons (in order to limit the observations to surface regions) requiring ultrahigh vacuum (UHV); electrons attached to the polymer surface could cause major modifications in the intrinsic structure of these surfaces.] As a result, most experimental data about surface tension of polymers is derived from measuring contact angles for drops of various solvents on the surface! The recent development of atomic force microscopes promises to provide new more direct information albeit on the micron scale rather than atomic (nanometer) scale.

We are using molecular dynamics simulations to predict the structure (reconstruction), surface tension, and other properties of polymer surfaces.

5.5.1 Polymer Surfaces

The crystalline forms of many polymers can be thought of as infinite molecular fibers closely packed (six neighbors to each fiber) as in Figure 18. This often leads to a rectangular unit cell (in the plane perpendicular to the chain axis) with two fibers per cell, where the chemical character of each fiber determines the orientation of the fiber about the chain axis. Often the fibers are oriented so that the strong interactions are along the b direction in Figure 18. In this case, the low energy surface should be the (100) surface (the yz plane, where z is along the chain axis and y is the b axis) indicated in Figure 18a. The other simple surface is (110) in Figure 19b. The (010) surface of Figure 19c breaks the strong bonds in the b direction and should be higher energy than (100) or (110). This is certainly true for Nylon 66 (where the hydrogen bonds are in the b direction) and for PEEK (due to electrostatic interactions in the b direction). It is less obvious which interactions should be strongest for PE and POM.

We are examining the structure (reconstruction), surface energies, and other properties for the [(100), (110), and (010)] surfaces of various polymers. The criteria for selecting these systems are:

1. that highly crystalline surfaces might eventually be prepared (for experimental test of the theory);
2. that the collection of systems includes cases with a variety of surface characters with widely varying surface energies (making experimental tests more significant); and
3. that they be useful models for industrial applications (making it more likely that data already exists and more likely that someone will be willing to do experiments).

An initial focus is on the reconstruction of the clean surfaces. Starting with the fully optimized structure for the bulk system, we calculate the system with free surfaces, allowing all atoms to relax. Effectively, these systems will have two-dimensional periodicity; however, for technical reasons (rapid convergence techniques for summing Coulomb interactions[15]), we treat them as three-dimensional cells where the slabs are separated by a distance D. Starting with the value of $D = D_0$ for which each slab is at its normal bulk separation, we increase D until the system fractures. The energy increase is just twice the surface energy. Orienting the stress, perpendicular to the chain axis, allows us to consider surfaces such as those in Figure 19.

Experimental values of surface tension are mostly derived from measurements of contact angles on surfaces. Unfortunately the experimental materials generally have surfaces that are rather amorphous leading to considerable uncertainties in the surface tension of crystalline surfaces.

For PE we calculate surface energies of[62]

> 101.6 dyne/cm for the (100) surface at 0K, and
> 104.6 dyne/cm for the (010) surface at 0K.

116

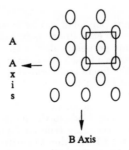

A

A ←
x
i
s

↓

B Axis

Figure 18. End view of closed-packed fibers along with unit cell notation.

Since the calculated lattice energy of PE is too high by 2%, we expect our calculations of surface energies to also be high by ~2%. Experimental data on PE suggest that the surface tension is 35 dyne/cm for amorphous PE and 66 dyne/cm for crystalline PE, both at 20C. We have not yet calculated the change in temperature difference of the surface energy from 0K to 300K, but expect that the surface energy at room temperature will be above 90 dyne/cm (again, we should be high by ~2%). The critical surface tension is related to the surface tension of the liquid that just wets the surface and hence need not correspond directly to the surface energy. However we find surface energies (at 0K) that are generally about twice the room temperature critical surface tension.

We should be able to obtain the mechanical properties (moduli) for the surface region by calculating the elastic constants as a function of slab thickness S (number of fiber layers). Because the surface fibers may rearrange to enhance the bonding at the surface, the surface elastic constants will change from that of the bulk and will depend upon the chemical characteristics of the fibers.

Calculated surface energies[67] for various polymer surfaces are reported in Table 11.

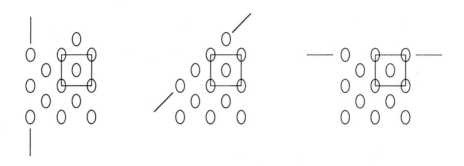

a. (100)　　　　　　b. (110)　　　　　　c. (010)

Figure 19. Illustration of crystal surfaces for various polymers.

Table 11. Calculated surface energies (ΔE/area) and experimental surface tension (γ_c). From reference 67.

Functional Unit	Molecule	ΔE/Area dyn/cm	γ_c Exper dyn/cm
$-CF_3$	$C_{12}F_{38}$	28	15
$-CF_2H$	$CF_2H-(CF_2)_{10}-CF_2H$	64	26
$-CFH_2$	$CF_2H_2-(CF_2)_{10}-CFH_2$	94	-
$-CF_2Cl$	$CF_2Cl-(CF_2)_{10}CF_2Cl$	41	~20
$-CH_3$	C_6H_{14}	64	30
$-CF$	$(CF)_N$	87	
	Graphite Fluoride	76	
$-CH$	$(CH)_n$	164	
$-CF_2-$	PTFE	29	23
$-CH_2-$	PE	102	66^a
$-CH-$	graphite	241	45
$-CH_2CHCl-$	PVC	255	43
$\left[-(CH_2)_2-O-\underset{O}{\overset{\|}{C}}-\text{⟨O⟩}-\underset{O}{\overset{\|}{C}}-O-\right]$	PET	230	44
$\left[-(CH_2)_6-\underset{\overset{\|}{O}}{\overset{H}{N}}-C-(CH_2)_4-\underset{\overset{\|}{H}}{\overset{O}{C}}-N-\right]$	Nylon 66	106	48

^{a}Extrapolated to crystal at 0K.

5.5.2 Polymer Melt-Solid Surface Interactions[67]

In addition to characterizing the structures and properties of the free surfaces, we will consider interactions with a second system. A useful simulation here is indicated in Figure 20. By comparing the energies of pure solid E_{cryst} in Figure 20a and pure liquid E_{solv} in Figure 20b with the E_{comb} of combined system in Figure 10c, we define the stabilization energy as

$$\Delta E_{stab} = E_{cryst} + E_{solv} - E_{comb} \qquad (15)$$

Some typical results are in Table 12. Such quantities will be useful for discussing issues of wettability, adhesion, and compatability. For systems such as PEEK, nylon, and epoxies we expect that the surface energies of different surfaces will change order as vacuum or hydrophobic liquids are replaced by hydrophilic liquids.

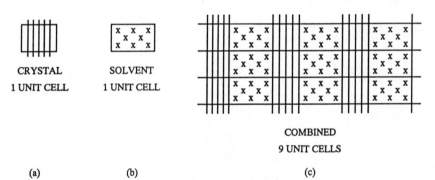

CRYSTAL   SOLVENT

1 UNIT CELL   1 UNIT CELL

COMBINED

9 UNIT CELLS

(a)   (b)   (c)

Figure 20. Illustration of calculations for surface-melt interactions.

Table 12. Calculated stabilization energies E_{stab} (erg/cm^2) from equation (15). (Reference 67.)

Solid Surface	Solvent	Theory	Experiment
Graphite[a]	CCl_4	115	115
$(CH)_n$	CCl_4	108	
$(CF)_n$	CCl_4	57	35
Graphite[a]	CH_2Cl_2	98	
$(CF)_n$	CH_2Cl_2	41	
Graphite[a]	benzene	138	

[a]Using reference 40.

5.5.3 Characterization of the Amorphous-Crystalline Interface in Polyolefins: A Combined Experimental and Modelling Approach[69]

The performance limits of many semicrystalline polymers, like high density polyethylene (HDPE), low density polyethylene (LDPE) or linear low density polyethylene (LLDPE), are dictated by their two-phase, amorphous-crystalline structure. For example, mechanical properties depend strongly on the nature of the interface between the amorphous and crystalline phases. In many cases the issue of two-phase morphology reduces in practice to the amount of two-phase interface. Much of the current catalyst research by polyethylene producers is directed toward design of new polyethylenes with controlled branching. But the full benefits of the designed polyethylenes can only be realized with a detailed knowledge of the interfacial structure associated with the inevitable semicrystalline structure of the new polymers.

The importance of two-phase structure in polyethylene has long been recognized, and in the absence of any detailed information on the interface, addressed primarily through trial-and-error variations in synthesis or processing. Now however, refined techniques in analytical chemistry, especially spectroscopy for interface analysis,[70,71] can be coupled with recent developments in computer modeling, specifically new algorithms designed for simulations of million-atom systems, to fully characterize the amorphous-crystal interface in polyethylenes. Such full characterization will pin-point the influence of branching in the novel polyethylenes on interfacial structure, and indicate how to control the structure through modifications in branching. Control of the amorphous-crystalline structure will result in extended life in current applications, such as better environmental stress crack resistance for HDPE pipe, and in improved performance in developing applications, such as polyolefins as matrix resins in filled systems for commercial construction.

Progress in this two-phase characterization depends critically on experimental measurement of interface features. Efforts are underway with MSC industrial partners to extend the expertise in analytic spectroscopy for two-phase oligomeric systems to polymer systems, preliminary analytical data on the effects of branching on crystallization already suggest a set of definitive experiments.

5.6 Modulus-Temperature-Time Behavior[68]

The thermoelastic properties of polymers typify the applications which motivate these developments in atomistic simulations. For polymers and composites the modulus-temperature-time behavior (e.g., the glass temperature, T_g) is at the heart of most applications. Here one needs to predict how T_g changes with blending, crosslinking, copolymerization, fiber preparation etc. Since the materials of interest are amorphous or (worse) partially crystalline and heterogeneous (composites), about 1 million atoms per periodic supercell may be required to provide a realistic model of the real polymer properties. The plan here is to:

1. develop reliable atomic level simulation procedures for predicting the moduli as a function of temperature and as a function of frequency of applied loads;
2. validate this technology by applying these procedures to several polymers, copolymers, and blends for which there is reliable data;
3. test this technology by using these procedures to predict the modulus-temperature-frequency characteristics for new cases that could be checked experimentally;
4. extract from the above studies an atomic level understanding of how T_g is related to molecular structure, packing, force fields (e.g., local stiffness), external stresses, molecular weight, void distribution, dilute impurities, solvents, etc.

The goal here is to attain a level of understanding that would permit the engineer to design polymers having a prescribed heat distortion behavior. Here the need is first a qualitative understanding in order to reason what to try and second the quantitative simulation tools for carrying out computer experiments to optimize the design. Only after the computer design need one carry out the expensive experiments of testing and verification. This should provide substantial savings from reduced experimental tests and also should allow the designer to consider completely new strategies.

6. Applications to Ceramics and Semiconductors

We are developing force fields for ceramics to use in predicting structures and energetics. The initial focus for ceramics is on Si_3N_4, SiO_2, and silicon oxycarbide. For semiconductivity it is group IV and III-V systems.[46]

6.1 Silicon Nitride Force Field[54]
6.1a Introduction

Silicon nitride has long been recognized as a promising high temperature structural ceramic for use in diesel engines, industrial heat exchangers, and gas turbines, to name but a few of its potential applications. Good thermal stress, oxidation, corrosion, and erosion resistance allow silicon nitride to operate in high temperature, high stress environments without the need for cooling. This is a distinct advantage over the materials currently used in these applications (alloys of chromium, cobalt, nickel, and tungsten) which require constant cooling to avoid melting. The reduced heat loss to cooling in engines insulates with silicon nitride increased the efficiency of these engines. Efforts to implement the use of silicon nitride in such applications, have been frustrated however, by problems with brittleness (often leading to catastrophic fracture), surface adhesion, and processing of the raw material into useful forms. This last problem has given rise to a family of silicon nitride based materials which utilize different sintering additives and hot pressing conditions to produce compounds with varying densities, $\alpha : \beta$ ratios, and intergranular structures. Most of these materials have strength and stress resistances far below the theoretical limits.

The most commonly used sintering additives (silicates and metal oxides) tend to conglomerate at intergranular boundaries. Softening of these intergranular glassy phases has been indicted as the main cause of creep and subsequent thermal breakdown in commercial materials. An atomic-level understanding of the behavior of different additives at the grain boundaries would facilitate prediction of the tribological properties of new and potentially more desirable silicon nitride based materials. As a first step towards modeling the relevant surfaces and interfaces of commercial silicon nitrides, we have studied the basic physical properties of the α and β phases of Si_3N_4 systems. Since almost no data is available for the pure systems, we view our results as benchmarks against which future forms of silicon nitride can be compared as processing technology advances.

6.1b Calculations

A force field for the alpha and beta forms of silicon nitride was derived using the HBFF/SVD technique combined with *ab initio* calculations on $N(SiH_3)_3$ and $Si(NH_2)_4$ clusters which model the nitrogen and silicon centers of the bulk material. This force field was used to predict crystal structures, lattice expansion parameters, elastic constants, phonons, and thermodynamic properties for pure α and $\beta - Si_3N_4$. Experimental measurements on these important physical constants are lacking, and our calculations provide the first reliable source of data on the fundamental properties of α and β-Si_3N_4.

We find good agreement of our calculated values with experimental properties such as crystal structures, lattice expansion parameters, and the thermodynamic functions C_v and S. We predicted phonon dispersions (Figure 21) and elastic constants but there are no sufficiently accurate measurements with which to compare. The expansion of the lattice parameters with temperature is shown in Figure 22 (for the α phase) where we see that the theoretical lattice expansion accurately tracks experiment.

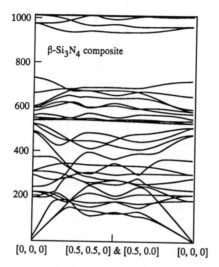

Figure 21. Phonon dispersion curves for $\beta - Si_3N_4$. (From reference 54.)

122

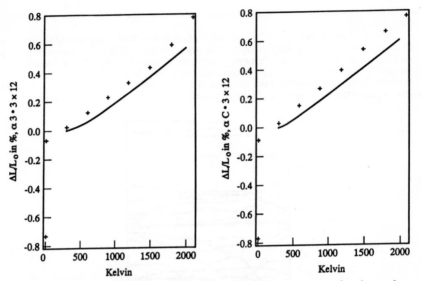

Figure 22. Predicted temperature dependence of lattice parameters for the α phase of Si_3N_4. The solid line is an interpolation of the experimental curve and the individual points were calculated by performing molecular dynamics for 20 ps on extended structures and averaging the cell parameters over the last 15 ps. (From reference 54.)

Calculating the vibrational modes for 1000 points in the Brillouin zone and then using the quantum statistical formulae for the thermodynamic properties leads to the results (solid line) in Figure 23 for specific heat (C_v) and entropy (S). There are two sets of data to compare with: *(i)* the JANAF Thermochemical Tables and *(ii)* the handbook Thermoproperties of Individual Substances (TPIS). Our calculations are in nearly exact agreement with TPIS for temperatures up to ~300K but differ substantially with JANAF. We have not yet examined the sources of experimental data to see which is more reliable. At higher temperatures the S and C_v from theory lag below the experimental results. This results from the quasi-harmonic approximation, implicit in our calculations (which uses only the minimized structure at 0K). Using canonical dynamics, the thermodynamic properties will reflect the sampling of the anharmonic regions of our force field and will lead to higher values of S and C_v. We are testing this.

Of particular interest is our calculated result that $\beta - Si_3N_4$ has a lower free energy than $\alpha - Si_3N_4$ over the entire temperature range sampled $(0 - 4200K)$, indicating that $\beta - Si_3N_4$ is the thermodynamically favored structure. This result may account for the currently unexplained lack of a β to α transition. Experimentally, the α structure is observed exclusively up to 1500-1700K regardless of the method of formation (i.e. sintering, CVD, hot-pressing, etc.). In the presence of a liquid phase like molten silicate, a secondary reconstructive transformation from the α to β phases takes place. The transition has never been observed in the absence of a liquid phase, which acts as a solvent in which the $\alpha - Si_3N_4$ can dissolve and recrystallize as β. Likewise, the reverse transition from β to α has never been observed at any temperature. Previous Madelung energy calculations agree with our thermodynamic results that the β form is more stable at 293K, well below the observed transition temperature. These details can be reconciled only if the α phase is kinetically but not thermodynamically favored at low temperature. At the transition temperature, sufficient energy is available to surmount the barrier to the thermodynamically favored β form. The β to α transition does not occur because at low temperature where the α and β forms become equally stable thermodynamically (this is the limit of 0K in our calculations), there is not enough energy available to drive the reconstructive transformation, which requires full breakdown of the crystal lattice. Furthermore, the α to β transition does not occur at lower temperatures because solvents suitable for $\alpha - Si_3N_4$ do not melt until 1500-1700K. The calculated elastic constants are in Table 13.

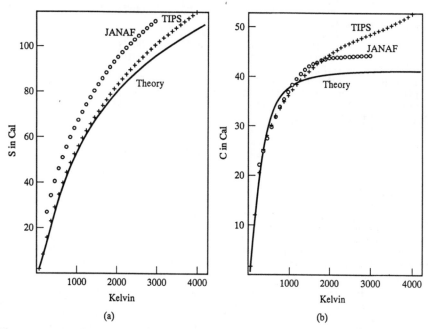

Figure 23. Predicted thermodynamic properties of $\alpha - Si_3N_4$. (a) entropy (S), (b) specific heat (C_v). The solid is from theory. (From reference 54.)

Table 13. Elastic constants and related quantities for Si_3N_4. (From reference 54).

	300K	Alpha 0K	300 K	Beta 0K
Stiffness Constants (GPa):				
C_{11}	426.854	436.745	439.172	447.541
C_{12}	156.903	175.416	181.848	214.550
C_{13}	152.039	176.016	149.909	165.140
C_{33}	473.394	508.784	556.983	580.212
C_{44}	126.780	132.434	114.380	115.002
C_{66}^a	134.785	130.628	135.930	116.507
Stiffness Constants (1/GPa):				
S_{11}	.00288	.00294	.00286	.00303
S_{12}	-.00083	-.00090	-.00010	-.00127
S_{13}	-.00066	-.00071	-.00050	-.00050
S_{33}	.00254	.00245	.00206	.00201
S_{44}	.00791	.00756	.00874	.00870
S_{66}^a	.00744	.00767	.00735	.00858
Compressibilities (GPa):				
β_v	.00406	.00370	.00376	.00353
β_a	.00140	.00133	.00136	.00126
β_c	.00122	.00104	.00106	.00101
Young's Modulus (GPa):				
5° to (0001)	392.287	406.133	479.079	490.725
$< 1\bar{2}12 >$	333.555	339.155	318.346	308.160
$< \bar{2}421 >$	345.265	340.082	346.209	327.546
Velocities of Sound (km/sec):				
5° to (0001)	.353	.359	.387	.392
$< 1\bar{2}12 >$	.325	.328	.316	.311
$< \bar{2}421 >$	.331	.329	.329	.320

6.1c Future Work

We are using our new force field to simulate the structure of amorphous silicon nitride by performing Monte Carlo calculations on systems of randomly oriented SiN_4 tetrahedra. This will allow extraction of distribution functions and structure factors necessary for calculating material properties and predicting scattering patterns. This understanding of the amorphous structure should provide insight into why the formation of $\alpha - Si_3N_4$ is favored over $\beta - Si_3N_4$ at low temperatures.

6.2 SiC, Silicon Oxycarbide[55-57]

High strength, tough, light materials can be achieved through careful design of a composite material. The combination of a very strong, but brittle fiber within a tough matrix can result in a composite with the best properties of each material. On the other hand, poor design of the composite might lead to a material enhancing the weaknesses of each material. For many years traditional fracture mechanics approaches failed to identify the important microstructural features which determined whether the composite material would exhibit the desired properties. The ultimate concern is to prevent a crack from propagating through the material, causing catastrophic failure. The analysis of crack propagation through a composite is complicated by several factors; first, the fibers or whiskers have compliances differing greatly from that of the matrix, second, the fiber/matrix interface affects crack propagation into and out of the fiber and may deflect the crack tip along the fiber, rather than through it. Finally, for fibers that are distributed inhomogeneously though the matrix, properties can be greatly anisotropic. A strong matrix-fiber interface results in cracks that propagate through the composite leading to catastrophic failure. A weak matrix-fiber interface on the other hand leads to several desirable effects. One effect of a weak interface is to blunt the crack tip. The stress intensity in front of a crack tip decreases as the crack tip radius of curvature increases. This slows or stops crack propagation. Additionally, cracks propagating into a weak interface will tend to be deflected along the interface, lengthening the path of the crack, dissipating energy, and moving the crack tip out of the plane of maximum stress. Original attempts to design tough composites incorrectly focused on strengthening the matix-fiber interface. Current understanding now directs correct engineering design of the interface; however, little is understood about the structure or properties of the interface. It is our goal to understand the structure-properties relationship of the interface to guide future composite design via computer/molecular mechanics experiments.

The bulk force fields are derived by optimizing force constants to fit the lattice parameters, the five key phonon curves (X and Γ points), and the three elastic constants. The resulting phonon dispersion[57] for $\alpha - S_iC$ is given in Figure 24. The LO-TO splitting at the Γ point is obtained by varying the ionic charges on the atoms. In contrast to the force fields developed for the zinc blende and group IV semiconductors, it was necessary to include van der Waals parameters in the optimization. The resulting force field will be coupled with force fields now being developed to model interface structures. We have now completed[55] a force field for studying amorphous silicon oxycarbide and are in the process of developing Monte Carlo procedures for predicting amorphous structures.[55]

126

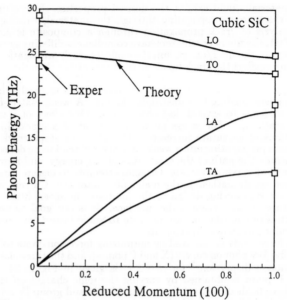

Figure 24. The predicted phonon dispersion curves for $\alpha - SiC$. (From reference 57.)

7. Conclusion

We summarized here some of the recent progress at the MSC/Caltech in developing new methodology for simulations of materials and for using these methods for industrially interesting applications. We believe that significant progress has been made and that the current technology is already useful to industry. However, there remains an enormous amount of research to develop the simulation methodologies required by current and future materials technologies. We believe that over the next ten years, theory and simulation will become an integral, critical part of industrial materials technology.

8. Acknowledgements

We are grateful for the government and industrial organizations that helped establish the MSC Resource Center in the Beckman Institute at Caltech.

Particularly noteworthy is AICD-DOE. This is the Advanced Industrial Concepts Division of DOE (before 1990 it was known as ECUT for Energy Conversion and Utilization Technologies) which was chartered to fund the research and development activities needed to bridge the gap between university research and industrial practise. They funded atomistic simulations for biotechnology (not reviewed here) that led to many of the advances and to the first collaborations with industry. In addition they provided the seed funding in 1990 for establishing the MSC. Jim Eberhardt of DOE and Minoo Dastoor of JPL get special credit for their early efforts and David Boron of DOE and Gene Petersen of NREL for later efforts.

Second, we would like to credit the reserach grants awarded over the years (1978-1990) from DARPA (Defense Advanced Research Projects Agency), ONR (Office of Naval Research), and AFOSR (Air Force Office of Sponsored Research) that enabled us to develop the infrastructure (state-of-the-art computers) required for tackling the industrially relevant materials problems. In particular the AFOSR program on fundamentals of tribology (L. Burgraf of AFOSR) intiated the developments leading to CMM, PS-GVB, and our work on ceramics.

Third, we must emphasize the importance of the continued funding from NSF (National Science Foundation, both the chemistry division and the division of materials research. The NSF funding (currently CHE-9010284, Dick Hildebrand) has allowed us to maintain an emphasis on the fundamentals of theory that has provided the advances now leading to industrial applications. In addition we are most grateful to NSF (Al Thayler) for funding the Grand Challenge Application Group (Goddard, Friesner, Taylor, Wang, and Jain) that is allowing us to accelerate the development of methods and software for utilizing high performance parallel computers in atomistic simulations.

Fourth, we would like to thank Arnold Beckman for establishing the Beckman Institute (headed by H. Gray) that has provided the space for the MSC and part of the funds for managing the MSC.

Fifth, we want to thank our industrial sponsors: BP America (Jim Burrington), Allied Signal (Mary Good), Asahi Glass (Kimihiko Sato), Asahi Chemical (Hiroaki Watanabe and Takahiro Komatsu), Xerox (Charles Duke), Chevron (Donald Paul), BF Goodrich (Ray Wissinger), Gore Corp. (Bob Henn), Nippon Steel, Shell Development Corp., and Dow Chemical for providing important funding and stimulation.

In addition there are many other current and past members of the MSC and the Goddard research group that contributed ideas leading to the work presented here.

9. References

1. F. W. Bobrowicz and W. A. Goddard III, in *Modern Theoretical Chemistry: Methods of Electronic Structure Theory*, H. F. Schaefer III, Ed. (Plenum Press, NY, 1977) Vol. 3, Chap. 4, pp. 79-127.
2. L. G. Yaffe and W. A. Goddard III, Phys. Rev. **A 13** 1682 (1976).
3. M. L. Steigerwald and W. A. Goddard III, J. Am. Chem. Soc. **106** 308 (1984).
4. A. Rappè and W. A. Goddard III, J. Am. Chem. Soc. **104** 448, 3287 (1982).
5. J. J. Low and W. A. Goddard III, J. Am. Chem. Soc. **108** 6115 (1986).
6. E. A. Carter and W. A. Goddard III, J. Catal. **112** 80 (1988).
7. M. Ringnalda, Y. Won, and F. Friesner, J. Chem. Phys. **92** 1163 (1990).
8. R. Friesner, J. Chem. Phys. **85** 1462 (1986); **86** 3522 (1987).
9. M. N. Ringnalda, M. Belhadj, and R. A. Friesner, J. Chem. Phys. **93** 3397 (1990). Y. Won, J-G. Lee, M. N. Ringnalda, and R. A. Friesner, J. Chem. Phys. **94** 8152 (1991). R. A. Friesner, Ann. Rev. Phys. Chem. **42** 341 (1991).
10. (a) J-M. Langlois, R. P. Muller, T. R. Coley, W. A. Goddard III, M. N. Ringnalda, Y. Won, and R. A. Friesner, J. Chem. Phys. **92** 7488 (1990). (b) R. P. Muller, J-M. Langlois, M. N. Ringnalda, R. A. Friesner, and W. A. Goddard, III, "A Generalized Direct Inversion of the Iterative Subspace Approache for Converging Generalized Valence Bond Wavefunctions (GVB-DIIS)," to be submitted.
11. J-M. Langlois, M. Ringnalda, and W. A. Goddard III, "Pseudospectral Methods for HF and GVB Calculations with Periodic Boundary Conditions," work in progress.
12. R. Friesner, H. H. Greeley, T. V. Russo, B. Martin, M. Ringnalda, J-M. Langlois, and W. A. Goddard III, work in progress.

128

13. C. F. Melius, B. D. Olafson, and W. A. Goddard III, Chem. Phys. Lett. **28** 457 (1974). C. F. Melius and W. A. Goddard III, Phys. Rev. **A 10** 1528 (1974). A. Redondo, W. A. Goddard, III, and T. C. McGill, Phys. Rev. **B 15** 5038 (1977). A. K. Rappè, T. A. Smedley, and W. A. Goddard III, J. Phys. Chem. **85** 1662 (1981).
14. P. J. Hay and W. R. Wadt, J. Chem. Phys. **82** 270, 284, 299 (1985).
15. N. Karasawa and W. A. Goddard III, J. Phys. Chem. **93** 7320 (1989).
16. J. Andzelm and E. Wimmer, J. Chem. Phys. **96** 1280 (1992).
17. X. J. Chen and F. Toigo, Solid State Comm. **79** 457 (1991).
18. S. Baroni and P. Giannozzi, Europhys. Lett. **17** 547 (1992).
19. R. Car and M. Parrinello, Phys. Rev. Lett. **55** 2471 (1985).
20. X.J. Chen, J-M. Langlois, and W. A. Goddard III, work in progress.
21. P. Weiner and P. Kollman, J. Comp. Chem. **2** 287 (1981). P. Weiner, et. al. J. Am. Chem. Soc. **106** 765 (1984).
22. B. Brooks, et al. J. Comp. Chem. **4** 187 (1983).
23. H. Q. Ding, N. Karasawa, and W. A. Goddard III, J. Chem. Phys. **97** 4309 (1992).
24. D. A. Tomalia, A. M. Naylor, and W. A. Goddard III, Angew. Chem. Int. Ed. Engl. **29** 138 (1990).
25. A. M. Naylor and W. A. Goddard III, *BioCatalysis and Bioimetrics, ACS Symposium Series 392*, J. D. Burrington and D. S. Clark, Eds. (American Chemical Society, Washington, DC, 1989), Chapter 6, pp. 65-87.
26. K. T. Lim, N. Karasawa, H. Q. Ding, and W. A. Goddard III, work in progress.
27. H-Q. Ding, N. Karasawa, and W. A. Goddard III, Chem. Phys. Lett., **193** 197 (1992).
28. H. Q. Ding, N. Karasawa, and W. A. Goddard III, Chem. Phys. Lett. **196** 6 (1992).
29. M. Tuckerman, B. J. Berne, and G. J. Martyna, J. Chem. Phys., **97**, 1990 (1992); see also J. Chem. Phys. **93** 1287 (1990), ibid **94** 1465, 6811 (1991).
30. J-M. Langlois, M. N. Ringnalda,R. F. Freisner and W. A. Goddard III, "Potential Derived Charges from Pseudospectral Hartree-Fock and Generalized Valence Bond Calculations" to be published.
31. A. K. Rappè and W. A. Goddard III, J. Phys. Chem. **95** 3358 (1991).
32. N. Karasawa, G. H. Miller, and W. A. Goddard III, "Charge Equilibration for Crystals," to be submitted.
33. A. K. Rappè and W. A. Goddard III, "Generalized Mulliken-Pauling Electronegativities. I. Main Group Elements" J. Phys. Chem, submitted for publication; A. K. Rappè and W. A. Goddard III, "Generalized Mulliken-Pauling Electronegativities. II. Transition Metals, Lanthaanides, Actinides, and Groups 2, 12, and 18" J. Phys. Chem., submitted for publication.
34. G. H. Miller, M. S. T. Bukowinski, R. Jean Loz and W. A. Goddard III, "Electronegativity Generation and the Dynamics of Ionic Materials" to be submitted.
35. J. J. Gerdy and W. A. Goddard III, unpublished.
36. S. L. Mayo, B. D. Olafson and W. A. Goddard III, J. Phys. Chem **94** 8897 (1990).
37. A. K. Rappè, C. J. Casewit, K. S. Colwell, W. A. Goddard III, and W. M. Skiff. "UFF, The Universal Rule Based Force Field for Molecular Dynamics Simulations of the Full Periodic Table," JACS, **114** 10024 (1992).

38. S. Dasgupta and W. A. Goddard III, J. Chem. Phys. **90** 7207 (1989)
39. T. E. Yamasaki, S. Dasgupta, and W. A. Goddard III, "Hessian Biased Force Fields: II The Singular Value Decomposition (SVD) Based Least Squares Method for Optimization and Analysis of Force Field Parameters," to be submitted.
40. W. A. Goddard and N. Karasawa, "Elastic Constants and Phonon States for Graphite; van der Waals Parameters for Carbon," J. Phys. Chem., submitted for publication.
42. M. Li and W. A. Goddard III, Phys. Rev. **B 40** 12155 (1989); M. Li and W. A. Goddard III, J. Chem. Phys., in press.
43. T. Yamasaki and W. A. Goddard III, "The Center-of-Mass Force Field for Organometallics," to be submitted.
44. N. Karasawa and W. A. Goddard III, "The Covalent Shell Model Force Field and Predicted Properties for Vinylidiene Difluoride Polymers," Macromolecules **25**, 7268 (1992); N. Karasawa, Ph.D. Thesis Applied Physics, Caltech, November, 1991
45. M. H. McAdon and W. A. Goddard III, J. Phys. Chem. 91, 2607 (1987); M. H. McAdon and W. A. Goddard III, J. Phys. Chem. **92** 1352 (1988).
46. C. Musgrave and W. A. Goddard III, "Force Fields for Group IV and III-V Semiconductors," in preparation.
47. This work is a collaboration involving T. Cagin of Molecular Simulations Inc., and N. Karasawa and W. A. Goddard III of MSC.
48. R. E. Donnelly and W. A. Goddard III, to be published; R. E. Donnelly, Ph.D. Thesis Chemistry, Caltech, February 1992; C. B. Musgrave, R. E. Donnelly, and W. A. Goddard, III, to be published.
49. S. Nosè, J. Chem. Phys. **81** 511 (1984).
50. M. Parrinello and A. Rahman, J. Appl. Phys. **52** 7182 (1981).
51. S. Nosè and M. L. Klein, Mol. Phys. **50** 1055 (1983).
52. H. C. Andersen, J. Chem. Phys. **72** 2384 (1980).
53. T. Cagin, N. Karasawa, S. Dasgupta, and W. A. Goddard III, "Computational Methods in Materials Science," MRS Symposium Series **278B** 61 (1992). T. Cagin, W. A. Goddard III, and M. L. Ary, J. Computational Polymer Science **I** 241 (1991).
54. J. Wendel and W. A. Goddard III, JCP **97** 5048 (1992).
55. B.L. Tsai, S. Dasgupta, J. J. Gerdy, C. B. Musgrave, and W. A. Goddard III, work in progress.
56. This work is in collaboration with Allied-Signal (C. Parker and S. Greiner).
57. C. B. Musgrave and W. A. Goddard III, work in progress.
58. A. Jain, N. Vaidehi, and W. A. Goddard III, work in progress.
59. A. Jain, N. Vaidehi, and G. Rodriquez, "A Fast Recursive Algorithm for Molecular Dynamics Simulations," J. Comp. Phys., submitted.
60. A. Mathiowetz, N. Karasawa, A. Jain, and W. A. Goddard III, "MD Calculations using NEIMO on all atoms of the Rhino Virus Capsid," work in progress. A. Mathiowetz, PhD Thesis Caltech, October 1992.
61. R. Friesner, R. Muller, and W. A. Goddard III, work in progress.
62. N. Karasawa, S. Dasgupta, and W. A. Goddard III, J. Phys. Chem. **95** 2260 (1991).
63. S. Dasgupta, N. Karasawa, and W. A. Goddard III work in progress.
64. W. Hammond, S. Dasgupta, and W. A. Goddard III, work in progress.

130

65. N. Karasawa, T. Cagin, and W. A. Goddard III, work in progress.
66. K. Smith, R. Stewart, S. Dasgupta, and W. A. Goddard III, work in progress.
67. N. Karasawa, T. Maekawa (Asahi Glass), T. Miyajima (Asahi Glass), and W. A. Goddard III, work in progress.
68. G. Gao, M. Belmares, N. Karasawa, S. Dasgupta, W. A. Goddard, III, R. Wissinger (BF Goodrich), D. White (BF Goodrich), P. Adriani, J. Schnitzer (BF Goodrich), work in progess.
69. J. Kerins (BP America), N. Karasawa and W. A. Goddard III
70. G. R. Stobe and W. Hagedorn, J. Polymer Science 16 1181 (1978).
71. R. Kitamaru, F. Horil, and S-H. Hyon, J. Polymer Science 15 821 (1977).

MOLECULAR MODELING AS A TOOL TO HELP DESIGN SELECTIVE ANTICHAGASIC DRUGS

Fulvia M.L.G. Stamato[1], Eduardo Horjales[2], Margot Paulino-Blumenfeld[3], Noriko Hikichi[3], Maria Hansz[3], Baldomero Oliva[4], Olle Nilsson[4] & O. Tapia[4]

[1] Departamento de Química, Universidade Federal de S.Carlos, C.P 676, 13560 S. Carlos, SP, Brasil
[2] Instituto de Física e Química de S.Carlos, Universidade de S.Paulo, C.P. 369, 13560 S.Carlos, SP, Brasil.
[3] Facultad de Quimica, Universidad de la Republica, C.C. 1157, 11600 Montevideo, Uruguay
[4] Physical Chemistry Department, University of Uppsala, Box 532, 75121 Uppsala, Sweden

1. INTRODUCTION

A variety of tropical diseases are produced by eucaryotic protozoa. Among them, trypanosomes include the causative agents of American trypanosomiasis or Chagas' disease (*Trypanosoma cruzi*), nagana (*T. brucei, T. vivax, T. congolense*) and african trypanosomiasis (sleep sickness; e.g. *T. rodhiense, T .gambiense*), while some species of plasmodes causes malaria (e.g. *Plasmodium falciparum, P. vivax*); related protozoa, such as leishmanias, provoke leishmaniosis (*Leishmania donovani*), oriental sore (*L. tropical*) and kala-azar (*L. donovani*).

Chagas' disease, endemic in Latin America, is caused by a pleomorphic organism which presents two phases in its vital cycle: one, in the mammalian host (man or domestic animals), the other, in the hematophagic insect. The acute phase of the affection, which occurs approximately 10-15 days after infection, may be lethal, but it generally evolves into a chronic phase with terminal illness characterized by extensive damage to heart muscle cells (myocardiopathy) or dilatation of the gullet and bowels nerves (mega-esophagus and megacolon). Effective drugs for the treatment of the chronic phase of American trypanosomiasis are not yet available. Those used in the acute phase and in congenital infection present undesirable side effects associated to their high toxicity. One of them, Nifurtimox, a nitrofuran derivative, is a fairly succesful antichagasic drugs; however, it is mutagenic in bacteria and may produce adverse effects in patients [1-3]. There is, then, an urgent need of new drugs for treating this disease.

Several strategies may be used to develop a new drug [4], such as empirical screening of chemical compounds, selective screening of known pharmaceutical agents with broad spectrum of activity, extraction from natural sources (plants or animal organs) and/or molecular modification. However, the development of a new drug demands high levels of scientific efforts and financial investiments. In the case of Chagas' disease, which is endemic in developing countries, the contribution coming from local scientific communities is highly desirable. It also opens a door to introduce new and advanced theoretical and experimental techniques contributing to development.

In the last years, there has been an increasing tendency to obtain new drugs through rational molecular design. This approach requires multidisciplinary collective effort of researchers from traditional and well established fields (such as Biochemistry, Chemistry, Molecular Biology, Immunology, Parasitology, Crystallography, etc.) and from recent advances in theoretically based simulation methods and computer technology: Computer-assisted Molecular Modeling. This work is concerned with the application of molecular modeling to the design of hopefully selective antichagasic drugs, via the identification of key structural differences between functionally related enzymes appearing in the parasite and mammalian host .

In spite of presenting striking structural and mechanistic similarities, Human Glutathione reductase (GR) and protozoal Trypanothione reductase (TpR) show a remarkable degree of mutually exclusive substrate specificity. This makes specific inhibition of the parasite enzyme in the presence of the host reductase an attractive possibility to be exploited in a rational design of selective antichagasic drugs [5,6]. Such studies, targeting the parasite reductase, requires knowledge of its three dimensional (3D) structure. Model building techniques were used to generate a 3D model for TpR from *T. congolense* [7]. Recently, crystal structures from *Crythidia fasciculata* were reported [8,9]. Docking studies have been carried out with the model TpR structure [7,8].

The paper is organized as follows. After a short description of design strategy in section 2, a brief presentation of the model building process of TpR from *T. congolense* is made in section 3. In section 4, comparisons between TpR model structure and the crystallographic ones for GR, LipDH and the newly reported crystal structure for TpR from *C. fasciculata* [8] are carried out. In section 5, the comparisons are focused on the active site, where substrate binding is examined. Possible sources of substrate specificity are analyzed with the collected structural information. In section 6, attempts at docking inhibitors of GR and TpR are described. Nitrofurans were selected. Some conjectures related to correlations between inhibitor binding site and the catalytic mechanism of the enzyme are given in the last section 7.

2. DESIGN STRATEGY

The enzymes studied here are trypanothione reductase (TpR), glutathione reductase (GR) and lipoamide dehydrogenase (LipDH). They belong to the family of FAD-dependent NAD(P)H-(disulfide)-oxido-reductases and are highly similar in both structure and mechanism.

Recent works aimed at designing selective antichagasic drugs [5,10,11] have focused on the identification of key differences between the parasite and mammalian host metabolisms with respect to the elimination of free radicals. It is well known [12,13] that nitroderivatives, such as nifurtimox, in contact with certain enzymatic systems of the parasite or the host, give rise to highly reactive oxyradical species, according to reactions 1-8:

$$RNO_2 + NAD(P)H \text{ ------ oxidoreductase -----} > RNO_2^{-\cdot} + NAD(P)^+ \quad (1)$$
$$RNO_2^{-\cdot} + O_2 \text{ ----------------------------------} > RNO_2 + O_2^{-\cdot} \quad (2)$$
$$2\ O_2^{-\cdot} + 2\ H^+ \text{ ----- superoxide dismutase ------} > H_2O_2 + O_2 \quad (3)$$
$$2\ H_2O_2 \text{ ---------------- catalase ------------------} > O_2 + 2H_2O \quad (4)$$
$$RNO_2 \text{ ---------------- nitroreductase ------------} > RNO_2^{-\cdot} \quad (5)$$
$$RNO_2^{-\cdot} + H_2O_2 \text{ -------------------------------} > OH^\cdot + OH^- + RNO_2 \quad (6)$$

The superoxide anion reduces ferric ions and the ferrous ions complete the Haber-Weiss reaction with hydrogen peroxide produced in the reaction catalyzed by superoxide dismutase:

$$H_2O_2 + Fe^{+2} \text{ --------------------------------> } OH \cdot + OH^- + Fe^{+3} \tag{7}$$

The superoxide anion inhibits catalase, favoring Fenton reaction:

$$O_2 \cdot^- + H_2O_2 \text{ --- ----------------------------> } OH \cdot + OH^- + O_2 \tag{8}$$

These equations summarize the reactions related with oxidative stress.

In mammalian cells, free radicals detoxification mechanisms involve the enzymatic couple glutathione reductase/glutathione peroxidase; the substrate is the oxidized form of glutathione (GSSG). In trypanosomatide parasites, trypanothione reductase (TpR) exerts that same function. The substrate is N^1,N^8-bis-(glutathionyl)-spermidine, known as trypanothione (TSST). The antitoxic action observed derives from a series of reactions in which the reduced form of these substrates (GSH or TSH) are involved. For instance, the reactions 9-11 may take place:

$$GSH \text{ (TSH)} + OH \cdot \text{ ------------------> } GS \cdot \text{ (TS} \cdot) + H_2O \tag{9}$$
$$GSH \text{ (TSH)} + HO_2 \cdot \text{ ----------------> } GS \cdot \text{ (TS} \cdot) + H_2O_2 \tag{10}$$
$$GS \cdot \text{ (TS} \cdot) + GS \cdot \text{ (TS} \cdot) \text{ -------------> } GSSG \text{ (TSST)} \tag{11}$$

GSH and TSH are consumed in those metabolic processes. Their concentration is kept at adequate levels, either by synthesis or by the reduction of GSSG or TSST catalyzed by GR or TpR, respectively, at the expenses of NADPH (reaction 12):

$$GSSG \text{ (TSST)} + H^+ + NADPH \text{ ----GR/TR----> } NADP^+ + 2 GSH \text{ (TSH)} \tag{12}$$

GR and TpR keep a high ratio of [GSH] / [GSSG] and [TSH] / TSST] in the cell [14,15]. In the parasite, glutathione participates in the detoxification mechanism, but here, there is no GR. It is TSH which directly reacts with GSSG to reduce it into GSH. This reaction maintains the physiological levels of glutathione in the parasite.

3. MODEL BUILDING OF A 3D STRUCTURE OF *T. CONGOLENSE* TRYPANOTHIONE REDUCTASE

3.1- Model building

The 3D structure of trypanothione reductase from *T. congolense* was built with a combination of techniques: sequence alignment, molecular graphics, energy minimization and molecular dynamics (MD). We briefly highlight each one of these steps. Additional details may be found in the original reference [7].

Some basic requirements must be fulfilled in order to allow for the construction of a model from sequence alignment. First, the aminoacids sequence of the target protein and the 3D structure of at least one homologous (reference) protein must be known. It is desirable to have at disposal additional related aminoacid sequences, with their 3D structures, in order to help model non conserved regions (insertions or deletions).

From this point of view, the dimeric enzyme trypanothione reductase is an ideal system. A number of aminoacid sequences for related enzymes are known, such as GR from human erythrocytes [16] and *E. coli* [17], *Azotobacter vinelandii* LipDH [18] and TpR from different sources: *T. congolense* [19] and *C. fasciculata* [20].

GR is closely related with TpR. The catalytic mechanism of GR is known in some detail [15,21]: this enzyme utilizes the isoalloxazine ring of FAD to transfer electrons from NADPH to a cysteine residue of the redox-active disulfide bridge, that may form a thiolate anion. Alternatively, as a quantum chemical study in progress in our laboratories shows, one excess electron can be trapped in the S-S bridge without breaking the bond, but increasing the interatomic distance and decreasing the stretching force constant. In a further step, the disulfide bridge of the substrate is reduced by the negatively charged protein S-S bridge.

In order to build the TpR model, the sequences of human erythrocytes GR [16] and *A. vinelandii* LipDH [18] were initially realigned (see Table 1). The realignment was made by superimposing their crystallographic structures. Regions with homologous sequences are structurally similar. Using this fact, we obtained a sequence alignment and determined those aminoacids belonging to conserved secondary structural elements (α-helices and β-sheets). Thereafter, the sequence of *T. congolense* TpR [19] was aligned, assuming that no insertions or deletions could appear in conserved secondary structures. The high sequence identity obtained with this restriction confirms the target protein as a member of the same family as GR and LipDH. The 3D structural differences found in this process are comparable in amount and type to those present between GR and LipDH. In fact, the procedure yields a 40.1% residues identity between TpR and GR, a 31.2% identity between TpR and LipDH and a 31.5% identity between GR and LipDH. Furthermore, GR and TpR are both reductases, while LipDH catalyses the reaction in the opposite direction. For this reason, GR was selected as the reference structure for TpR model building; LipDH was used as an auxiliary information source. The main chain in the crystallographic structure of *A. vinelandii* LipDH [22] may be superimposed to that of human erythrocytes GR [23] with a root mean square deviation less than 1.1 A for the Cα atoms in secondary structures, comprising more than 70% of each structure. This determines a core for the 3D structure of the family that is conserved and constitutes a basis for building a model of the target protein accurate enough to allow for automatic simulations (energy minimization, EM, or Molecular Dynamics, MD).

Starting from the crystallographic model of human erythrocytes GR, the main chain of *T. congolense* TpR was built by replacing the mutated aminoacids with the aid of the molecular graphics program FRODO [24] implemented in a Silicon Graphics Iris 4D-25G workstation. Insertions and deletions in GR/TpR alignments were much more frequent than in LipDH/TpR alignment, despite the higher aminoacid identity among the former. This fact allowed the use of information from the LipDH crystal structure to build some loops of TpR. Local alignment of LipDH/GR structures were run to overcome domain-domain rotations present in these structures. After this procedure, only insertions with at most four aminoacid remained. The changes in geometry, determined by insertions and deletions in the primary sequence, were then regularized with the aid of FRODO's subroutine REFI. Loop regions corresponding to non conserved residues in both alignments were built with the aid of FRODO's subroutine BONES, by choosing pieces of refined structures from PDB files which fitted best to the main chain geometric conditions. The side chain positions were initially altered manually and further optimized by local side chain torsional angles energy minimization, by using the subroutine FIT of the molecular graphics program

Table 1- Sequence alignement for Human erythrocites GR [16], *T.congolense* TpR[19] and *A.vinelandii* LipDH [18] used to build up the TpR model structure [7].

```
                    1        10        20        30        40      49
                                                           42
1  >Glut.R.    1 ACRQEPQPQGPPPAAGAVASYDYLVIGGGSGGLASARRAAEL GARAAVV
2  >Trip.R.         MSKAFDLVIIGAGSGGLEAGWNAATLYKKRVAVV
3  >Lip.D.          SQKFDVIVIGAGPGGYVAAIKSAQL GLKTALI

                   53              60        70        80
                                                          85
1  >Glut.R.   50 ESHK       LGGTCVNVGCVPKKVMWNTAVHSEFMHDHADY GFP
2  >Trip.R.      DVQTVHGPPFFAALGGTCVNVGCVPKKLMVTGAQYMDQLRESAGFGWEFD
3  >Lip.D.       EKYKGAAGA AALGGTCLNVGCIPSKALLDSSYKFHEAHESFKLH GIS

                   90       100       110       120       130
                                                122       134
1  >Glut.R.   89 SCEGKFNWRVIKEKRDAYVSRLNAIYQNNLTKSH IEIIRGHAAFTS
2  >Trip.R.      ASTIKANWKTLIAAKNAAVLD INKSYEDMFKDTEGLEFFLGWGALEQKNV
3  >Lip.D.       TGEVAIDVPTMIARKDQIVRNLTGGVAS LIKANGVTLFEGHGKLLAGKK

                        140       150       160       170       180
                        141                 164
1  >Glut.R.  134     DPKPTIEVSGKKYTAPHILIATGGMPSTPHESQIPGASLGITSDGF
2  >Trip.R.      VTVTEGADPKSKVK    ERLQAEHIIIATGSWPQMLK    IPGIEHCISSNEA
3  >Lip.D.       VEVTA ADGSSQVL    DTENVILASGSKPVEIPPAP      VDQDVIVDSTGA

                        190       200       210       220
                                            209
1  >Glut.R.  181 FQLEELPGRSVIVGAGYIAVEMAGILSAL   GSKTSLMIRHDKVLRSFD
2  >Trip.R.      FYLEEPPRRVLTVGGGFISVEFAGIFNAYKPVGGKVTLCYRNNPILRGFD
3  >Lip.D.       LDFQNVPGKLGVIGAGVIGLELGSVWARL   GAEVTVLEAMDRFLPAVD

                    230       240       250       260       270
                              239                           268
1  >Glut.R.  228 SMISTNCTEELENAGVEVLKFSQVKEVKKTLSGLEVSMVTAVPGRLPVMT
2  >Trip.R.      YTLTQELTKQLV  ANGIDIMTNENPSKIELNPDGSKHVTF      ESG
3  >Lip.D.       EQVAKEAQKILT  KQGLKILLGARVTGTEVKNKQVTVKFV      DAEGE

                    280       290       300       310       320
1  >Glut.R.  278 MIPDVDCLLWAIGRVPNTKDLSLNKLGIQTDDKGHI IVDEFQNTNVKGIY
2  >Trip.R.      KTLDVDVVMMAIGRLPRTGYLQLQTVGVNLTDKGAIQVDEFSRTNVPNIY
3  >Lip.D.       KSQAFDKLIVAVGRRPVTTDLIAADSGVTLDERGFIYVDDYCATSVPGVY

                    330       340       350       360       370
                                                358
1  >Glut.R.  328 AVGDVCGKALLTPVAIAAGRKLAHRLFEYKEDSKLDYNNIPTVVFSHPPI
2  >Trip.R.      AIGDVTGRIMLTPVAINEGASVVDTIFGSKP RKTDHTRVASAVFSIPPI
3  >Lip.D.       AIGDVVRGAMLAHKASEEGVVVAERIAGHKA  QMNYDLIPAVIYTHPEI

                    380       390       400       410       420
                              391                       414
1  >Glut.R.  378 GTVGLTEDEAIHKYGIENVKTYSTSFTPMYHAVTKRK TKCVMKMVCANK
2  >Trip.R.      GTCGLTEEEAAKSF EKVAVYSTCFTPLMHNISGSKYKKFVAKIITDHG
3  >Lip.D.       AGVGKTEQALKAEGVAINVGVFP  FAASGRAMAAND TAGFVKVIADAK

                    430       440       450       460       470
1  >Glut.R.  427 EEKVVGIHMQGLGCDEMLQGFAVAVKMGATKADFDNTVAIHPTSSEELVT
2  >Trip.R.      DGTVVGVHLLGDSSPEIIQAVGICMKLNAKISDFYNTIGVHPTSAEELCS
3  >Lip.D.       TDRVLGVHVIGPSAAELVQQGAIAMEFGTSAEDLGMMVFAHPALSEALHE

1  >Glut.R.  477 LR
2  >Trip.R.      MRTPSHYYIKGEKMETLPDSSL
3  >Lip.D.       AALAVSG
```

TOM [25] in order to relieve bad contacts - that should produce large distortions in simulations - without affecting the conserved main chain 3D structure. This step is essential for getting a good starting conformation for EM and MD. The initial model-built TpR dimer (MB structure) did not contain the NADPH co-factor. This was not present in the PDB GR 3D structure.

Some conclusions can be attained from the model building procedure:

1. the attribution of insertions and deletions plays a fundamental role in this process, as none of them should destroy secondary structures nor affect regions involved in the catalytic mechanism of the enzyme within certain limits. A model which fulfill these conditions has good odds to become a reliable one.

2. the use of informations from homologous structures and biochemical data is critical to avoid model building errors.

3. manual procedures, in spite of demanding a longer work time, make use of all the available information and may result in a better knowledge of the structure.

4. automatic model-building procedures which allows for manual intervention in key steps need to be developed.

5. local side chain torsional angle energy minimization is a good alternative to automatically overcome initial bad contacts. In this way, available information about the structure may be used to accept only models which are compatible with the available data.

3.2. Model refinement and construction of the TpR-NADPH complex.

The next step was the refinement of the model of TpR monomer A by MD simulations; to this end, GROMOS package [26, 27] was used. 150 steps of a steepest descent energy minimization were initially run (thus obtaining an EM-MB structure), followed by a 5 ps MD shake-up. A conformation was selected at each ps and an energy-weighed average structure (EM-MD) was thereafter generated. This 5 ps shake-up aimed at detecting structural regions more or less prone to deformation under the influence of thermal fluctuations and those that were stable. In order to test the sensitivity of different regions in TpR EM-MD model during the MD shake-up, different regions, as defined by Rice *et al* [28], were fitted to the original EM-MB structure with respect to the core $C\alpha$ atoms. Some of them showed a root mean square (rms) deviation of about 2.0 A (α-helices 7, 10 and 13), others presented rms values between 1.0-1.5 A (α-helices 8 and 9) and others, rms less than 1.0 A (α-helices 2 and 5). Thus, TpR structure is globally stable, though some regions are potentially more sensitive to conformational variations than others in the course of a longer MD simulation.

The complex of NADPH with TpR was then built by first docking NADPH to GR, imposing constraint distances described in the literature [29]. This structure was used as a template to dock NADPH in TpR. This putative complex NADPH- TpR model has been further refined by energy minimization technique [26]. First, 500 steps of EM were run by imposing the restriction of a fixed distance (3.9 A) between FAD and NADPH and keeping the protein atoms also fixed. Then, 500 EM steps keeping NADPH and FAD fixed and allowing the protein to relax freely. Finally, 500 EM steps without any constraints. The final structure corresponds to our model of TpR monomer A.

4. STRUCTURAL COMPARISONS: TpR MODEL AND CRYSTAL GR, LipDH AND TpR.

4.1- General features of TpR model, GR and LipDH   3D structures.

The model for the TpR-NADPH monomer A described above preserves the secondary structure elements of the starting X-ray structure of GR and most of the interactions found in the GR-NADPH model.

Each subunit in these enzymes presents four different domains:
i) the N-terminal FAD binding domain;
ii) the NADPH binding domain;
iii) the central domain and
iv) a C-terminal interface domain.

The prostethic group, FAD, and the co-factor, NADPH, are bound in extended conformations (Figure 1). The nicotinamide moiety makes its most extensive contacts with the flavin [23]. Both of them make other contacts mainly with residues of their respective domains, although some residues of other domains are also involved in stabilizing interactions. For instance, the isoalloxazine ring of FAD is tightly bound through its N_3 atom to His467' of the other subunit, while there is a strong hydrogen bond between the O_2'N of the nicotinamide mononucleotide portion of NADPH and the carbonyl group of Leu337 [23].

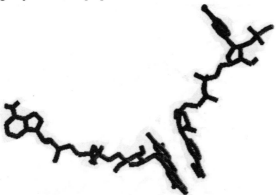

Figure 1- FAD   and NAD(P)H extended conformations in GR, TpR and LipDH.

The active site is placed in the interface between the two subunits. Residues from both of the subunits contribute to form it. The substrate binding site is placed in a deep crevice, separated from the NADPH binding site by the redox-active disulfide bridge of the enzyme and the FMN moiety of FAD. The topography described is basically the same in   these three enzymes. The active sites of GR and LipDH were preserved in the TpR model (Figure 2). During the 5 ps of MD, this region fluctuates without changing the spatial disposition of the catalytically important residues. The nicotinamide and isoalloxazine rings planes are almost parallel, with a mean distance of 3.57 A between C_4 of NADPH and N_5 of FAD.

The residues   of FAD's binding site in GR are highly conserved in TpR, with only two significant differences with respect to the corresponding site in LipDH. In GR and TpR, there is a hydrogen bond between the O_4 atom of the isoalloxazine ring of FAD and the N_Z atom of Lys66/GR (Lys59/TpR). Lys67/GR (Lys60/TpR) is

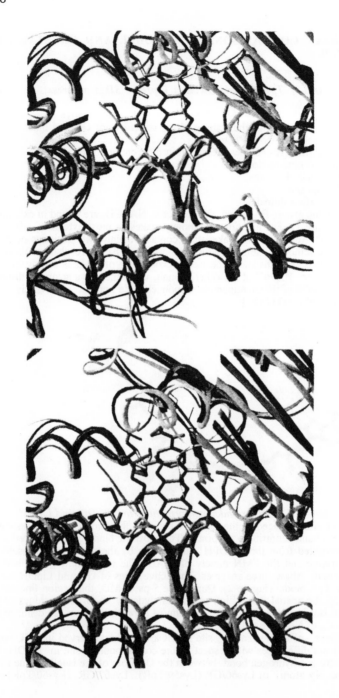

Figure 2- Stereoscopic view of the superposed backbones of GR, TpR and LipDH active sites. FAD, NADPH and TSST are also shown. (Dark grey, TpR model; medium grey, thick ribbons, TpR crystal; medium grey, thin ribbons, GR; light grey, LipDH)

involved in substrate binding and is directed towards the channel connecting the two subunits of GR (TpR) dimer. In LipDH, the residue 66 is an alanine and the residue hydrogen-bonded to O_4 now is Lys67, building up a much weaker hydrogen bond. This change may be related to differences in substrate specificity and reverse reactivity of LipDH, favoring the binding of the oxidized form of NAD (NAD^+) instead of NADPH as in GR and TpR. Another difference between the model structure of TpR and that of GR and LipDH refers to the interactions established by the O_3^* atom of the ribose in FAD. For GR and LipDH, this atom bears a strong hydrogen bond (2.8 A) to Glu50 in both cases. In TpR, this residue is an Asp, 3.3 A apart from O_3^*. Next to it, in TpR, there is a residue, His 54, placed in an insertion segment in relation to GR. We have studied the possibility of maintaining the interactions of these residues (Asp50 and His54), by building another strong hydrogen bond (2.8 A) between the O_3^* atom of FAD and His 54. However, the energy minimization and MD simulations showed that the energetically favored conformation corresponds to an interaction between His54 and the solvent, keeping the ribose atom hydrogen-bonded only to Asp50, like in the parent structures. For the binding site of NADPH, the interactions with the adenine moiety described for GR are all preserved in our TpR model. The other moieties of NADPH, *e.g.* ribose, 2'-phosphate, A-phosphate, N-phosphate and nicotinamide ring, in the TpR-NADPH model, contains most of the interactions described in the X-ray structure of the GR/NADPH complex. The polar interactions between TpR and FAD are practically the same as for GR, as well as the polar interactions between TpR-NADPH and GR-NADPH [7]. Insertions and deletions occurred only at the surface of the enzyme.

As expected for this type of model building, the overall 3D structure of GR is highly preserved in TpR (Figure 3). The dimeric structure of GR is symmetric and so is our TpR model. All the four domains of GR and TpR subunit A can be separately superimposed onto their counterparts in subunit B and also onto the corresponding one in the homologous enzyme. The same situation does not hold true when the Cα backbone of TpR model and LipDH crystal structures are superimposed as described above (Figure 4). This fact arises from a change in the relative orientation of the four domains of LipDH [30], determined by a considerable reorientation in the tertiary structure of the interface domain. Thus, taking the FAD domains of TpR model and LipDH crystal structure as a reference, after superposition, the other domains are rotated by several degrees. The substrate binding site of LipDH is more closed than the corresponding one in the TpR model-built structure. This fact may be traced back to the above mentioned rotation of domains in LipDH and also to the presence of an additional short tail at its C-terminal, that narrows the substrate binding site.

Important differences in non loop regions between the crystallographic model of GR and model-built TpR structure refers to the binding site of the substrates, which may account for the different substrate specificity (see below).

4.2- Comparison of TpR-model and crystallographic structures.

Between the   *C. fasciculata* and *T. congolense* aminoacid sequences there is an overall identity of 68%. This value can be compared with the identity of 40.1% obtained in the alignment of TpR and GR, thereby showing a higher degree of similarity, which should result in a closer resemblance between the two TpR structures.

The superposition of the dimeric model-built and X-ray structures of TpR [8] shows an almost perfect fit between the   subunits A of each dimer and a poorer agreement in subunit B, if the superposition is performed on subunit A (Figure 5). This is caused by a small (2.7^0) rotation in the relative orientation of the two monomers.

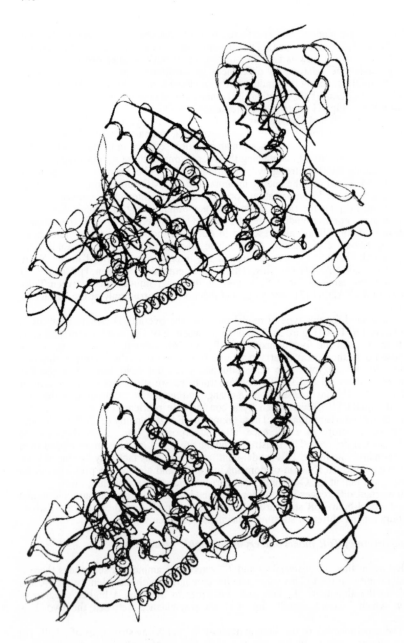

Figure 3- Stereoscopic view of the superposed backbones of GR X-ray structure[23] and TpR model-built structure [7]. (Dark grey, TpR; light grey, GR)

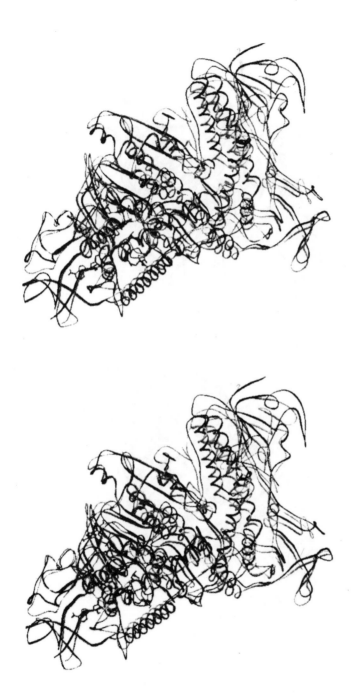

Figure 4- Stereoscopic view of the superposed backbones of LipDH X-ray structure [22] and TpR model-built structure [7]. (Dark grey, TpR; light grey, LipDH)

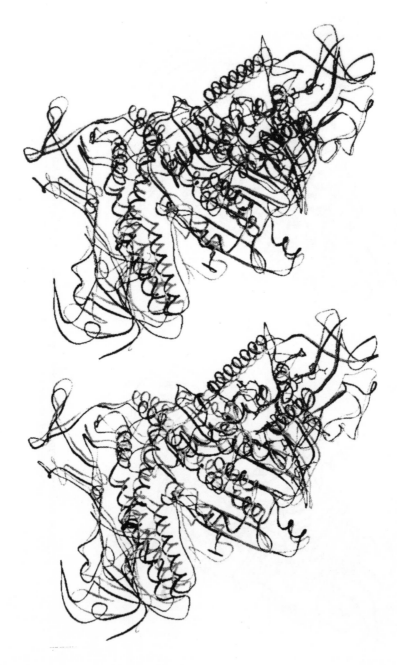

Figure 5- Stereoscopic view of the superposed backbones of TpR X-ray [8] and model-built [7] structures. (Dark grey, TpR model; light grey, TpR crystal)

In TpR X-ray structure, a 4° rotation of FAD, NADPH and central domains away from the interface domain of the other monomer was determined [8,30], which could be related to the above mentioned monomer rotation.

The active site of the X-ray structure is slightly more open than that of the model-built one. This result may be attributed to an interdomain rearrangement in the TpR X-ray structure, which, of course, is not present in the model. The largest difference between these two independent structures is found at the surface of the enzyme, in a region comprising residues 129 to 149 and constituted by two loops separated by a β–strand, which is formed by 4 aminoacid residues. In the model structure, this region was constructed by following the non-refined X-ray coordinates of *A. vinelandii* kindly provided by Prof. Wim Hol prior to publication. However, after refinement of the X-ray structure [22], it was found that this loop has the opposite direction to that previously assumed. This caused the difference in the model. The new orientation taken by the loop in the refined coordinates of LipDH happens to be similar in TpR X-ray structure.

Another difference between the two structures is found at the C-terminus. This portion of the molecule was not built in our TpR model, since in TpR it is 18 residues longer than in both GR and LipDH sequences. In the X-ray TpR structure [8], only the last 6 residues of the C-terminus were not included.

After these and other comparisons we can consider the unbiased model building of TpR from *T. congolense* as fairly successful. We report now the work carried out with this 3D structure.

5. ACTIVE SITES AND SUBSTRATE BINDING SITES COMPARISONS

Glutathione reductase presents a ping-pong mechanism under physiological conditions. First binds NADPH to the enzyme. Two electrons and one proton are left behind when the oxidized co-factor leaves the enzyme. Thereafter, the substrate - oxidized glutathione (GSSG) - binds to the reduced enzyme. The substrate is then reduced leaving two molecules of glutathione (GSH). The overall reaction corresponds to a transposed hydride transfer. The active site contains two differently located zones: one for NADPH nicotinamide ring and the other for GSSG. Pai and Schulz [21] have thoroughly studied the mechanism of GR. In what follows, we draw from their work the fundamental facts. Some alternatives are proposed that affect somewhat this mechanism.

5.1- Active sites topography

The active site of GR is located at the interface between the two monomers and is constituted by FAD and some aminoacid residues around the nicotinamide ring: an ionic pair, made up by Lys66 and Glu201, that appears to assist the initial step of the catalytic reaction; Tyr197 is placed nearby to the active site. This residue is flexible; in the absence of NADPH, the nicotinamide binding site is closed by a movement of its side chain, thus avoiding the solvent to come in contact with the electron transfer system, namely, the isoalloxaxine ring of FAD. We assume the role of this residue to be fundamental in the sense that it contributes to conform the active site surface which must be complementary in form to the transition structure involved in the electron transfer processes. If this hypothesis is correct, it can be predicted that any mutation of this residue by one having smaller surface will impair the catalytic efficiency of the enzyme.

In the substrate active site, Cys58 and Cys63 are involved in a redox active disulfide bridge. The reduced enzyme most likely contains one electron on this bridge (without bond breaking, but with small force constant and larger equilibrium distance) and the second electron resides in the isoalloxazine ring. This proposal is partly based upon *ab initio* molecular orbital calculations on the moieties participating in the electron transfer.

The substrate may form a charge transfer complex with the reduced S-S bridge; a proton relay system, already described by Pai and Schulz, constituted by residues His467' and Glu472' of the second subunit most likely provides the proton needed to reduce the disulfide bridge of the oxidized substrate in the final step of the catalytic mechanism of the enzyme.

The GSSG active site in GR has a Tyr114 that stacks between GS moieties. From our viewpoint, this residue seems to play a role similar to the Tyr197 at the nicotinamide binding site, namely, it contributes to close the shape of the active site in order to achieve complementarity to the transition structure of the interconversion step catalyzed by the enzyme. Mutations of this residue may impair the catalytic activity if there is no surface conservation.

The same situation is observed at the TpR active site, but there the residues numbering is altered due to the insertions and deletions; the S-S bridge links Cys51 to Cys56; the proton relay system is composed by His460' and Glu465' and the ionic pair is made up by Lys59 and Glu201. The role of Tyr197 in GR is now played by Phe197 and Tyr114 becomes Tyr110 in TpR.

The detailed knowledge of the substrate binding site of both TpR and GR enzymes is an essential requirement to design specific inhibitors for TpR. Thus, the interactions established between these enzymes and its natural substrates and known inhibitors of GR were studied in this work by molecular modeling techniques.

5.2- Substrate and inhibitor models.

Oxidized glutathione (GSSG), the natural substrate of GR, is conformed by two halves of L-glutamyl-(γ-cysteinyl- glycine), known as GSH , bounded by a disulfide bridge [31]. Considering the covalent structure alone, GSSG has an internal twofold symmetry axis and could in principle assume a symmetric conformation when bound to a symmetric dimeric structure like GR. However, GSSG binds to an asymmetric site at the dimer interface of glutathione reductase. Therefore, it is necessary to distinguish between the two halves of GSSG, which are designated as GS-I and GS-II, where GS-I is the half which is covalently linked to the enzyme during catalysis. Accordingly, the aminoacid residues of GSSG are named Glu-I, Cys-I, Gly-I and Glu-II, Cys-II and Glu-II.

Oxidized trypanothione (TSST, N^1,N^8-bis-(glutathionyl)-spermidine) is an unique analogue of glutathione; here, two glutathione tripeptide moieties are covalently linked in the glycine extremity via a spermidine bridge [32]. Unlike GSSG, reducing the disulfide bridge of TSST does not break the molecule in two, since the spermidine bridge is still connecting them.

The X-ray structures of the GR complexed with GSSG has not yet been deposited at the Protein Data Bank, although the relative orientation and contacts GSSG makes at GR active site were described [29]. In order to model GSSG and TSST, the molecular mechanics method of Allinger (MM2) [33] was adapted to calculate polypeptide structures by generating a new series of parameters named MM2G [34]. Some parameters of this first generation force field were modified, leading to a better agreement between the geometries obtained with this method for some peptides and the crystallographic ones. In particular, the root mean square deviation for glutathione decreased to 0.45 A , similar to values obtained with other procedures [35]. A model for the structure of trypanothione was also obtained by the same molecular mechanics procedure [36], by initially linking the two carboxyl glycine groups of GSSG with an extended spermidine bridge.

The model GSSG was docked into the 3D structure of GR until the situation described in the literature for the GR-GSSG complex [29] was reproduced. This structure was then used as a template to build the TSST-TpR complex. Alternatively, GSSG was docked at TpR's active site and TSST was docked at GR active site. Four different substrate-enzyme complexes were then generated: TpR-TSST, GR-GSSG, TpR-GSSG and GR-TSST [36], which allows for a comparison of the substrate binding site in both enzymes and may provide clues to understand the exclusive substrate specificity of the enzymes.

After the preceding work was completed, a crystallographic structure of GR complexed with an inhibitor (GSSG retro) was made available through PDB [37]. This is also examined here, to check up the conclusions obtained from the present docking studies.

On the other hand, it is known that GR preparations from different sources are inhibited by nitro compounds [13]. Several nitrofuran derivatives have been synthesized [38] and their properties determined [39, 40]. Among them, 1-(5-nitro-furfuryliden-amino)-pyrazole (NF-pyrazole) has proved to be more effective than nifurtimox as inhibitor of yeast and mammalian GR and was used in our work as a model compound to study the putative inhibition site of the enzyme by nitrofuran derivatives. The binding site for nitrofurans in GR is not known. The binding of glutathione together with some related molecules and two redox compounds (safranine and menadione) to crystals of GR was investigated by X-ray crystallography at 3 A resolution [31]. The glutathione binding site was well characterized and all glutathione conjugates and derivatives studied in that experimental work showed binding dominated by interactions at this site. CoASSG (a mixed disulfide between coenzyme A and glutathione), in addition to the binding at the GSSG site, also binds at the adenine site of NADPH. A third binding site, in a large cavity of GR located at the interface of the two subunits, between Phe78/Phe78' and His75/His75', was observed for safranine and menadione. This site was called regulative or interface binding site.

Three possible substrate and inhibitors binding sites in GR were then put in evidence by X-ray studies: the substrate binding site, the NADPH binding site and the interface binding site. We have therefore examined the binding of NF-pyrazole at these three sites (see below, sect. 6.2.3). The program TOM was used to dock the inhibitor molecule to the X-ray structures of GR and LipDH and the model-built structure of TpR.

5.3 Substrate binding

GSSG binding site in GR is formed up by various parts of the chains of both subunits. On binding, GSSG makes connections to several aminoacid residues, which altogether have four positive (Arg37, Arg38, Arg347 and Arg478') and two negative charges (Glu472' and Glu473'), which as a whole compensates the charges of the carboxylate groups of GSSG. The X-ray study of GR-GSSG complex [29,31] shows that the α-carboxylate group of Glu-I interacts with Arg347 and a bound water molecule; the amino group is directed towards the solvent bulk. Cys-I S-atom binds within van der Waals distance of Cys58 SG and His467'. Its backbone is also in contact with the end of the side chain of Tyr114, forcing it to move nearly 1.0 A away from the crystallographic position in the apo-enzyme. There are five residues contacting Cys-I; none of them are exchanged in TpR, thus suggesting Cys-I location is conserved. Gly-I is fairly well buried; its carboxylate makes contact with Arg37. The pocket of Gly-I can be opened by the side chain movement of Tyr114, the amide group of Gly-I pointing towards the hydroxyl group of Tyr114, but being too far away to interact strongly. On the other hand, Glu-II makes direct contacts with residues from the other subunit of the enzyme only. Its α-amino group is held by three hydrogen bond acceptors groups: Thr469' O, Glu472' OE_1 and Glu473' OE_2. The α-carboxylate receives a hydrogen bond from Met406' N and from two solvent molecules which contact the enzyme at Lys67 and Ser470', respectively. Cys-II makes weak contacts with the end of the side chains of Leu110, Tyr114 and His467' and seems to be held in place primarily by the disulfide bond to Cys-I. The electron density distribution of Gly-II shows that this residue is not well defined upon binding The amide group has a weak interaction with the O_Z atom of Tyr114.

The same situation was reproduced in our model-built GR-GSSG complex (Figure 6). It is expected the glutathione molecule to be held in place at its GR binding site mainly because of the interactions set up with the array of positively charged residues placed nearby to the region occupied by Gly-I and Gly-II (Arg37, Arg38, Arg347 and Arg478'), which helps to stabilize the negative charges of the carboxylate groups of the substrate. Moreover, Tyr114 is in a correct position to complete the active site inner surface complementary in shape to the transition structure of the reaction catalyzed by GR.

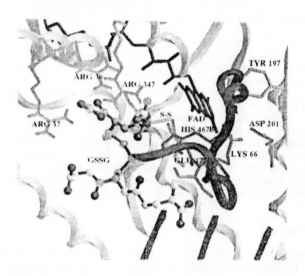

Figure 6- GSSG docked in GR active site.

Due to the mutations observed in its substrate binding site and to conformational differences in its longer C-terminal, the distribution of the TpR charged residues in this region has been drastically altered with respect to GR and contains a series of neutral or negative aminoacid residues: Glu17, the TpR counterpart of GR

Ala34; Trp20 (Arg37 in GR), Asn21 (Arg38 in GR) and Ala342 (Arg347 in GR). Arg471', which is conserved in GR, is no longer pointing towards the substrate binding region due to the existence of the longer C-terminus in TpR.

Trypanothione, TSST, presents drastic changes in the glycyl moiety of the glutathione parent structure. One carboxylate in each molecule GS-I and GS-II is neutralized by covalently binding to spermidine to form TSST; the carboxylates from the glutamyl moieties are preserved. One of these carboxylates and the positively charged nitrogen atom of spermidine bind in the above mentioned region of TpR corresponding to the GSSG binding site in GR: the carboxylate group of Glu17 would interact with the positively charged nitrogen group of the spermidine bridge. Note that Kuriyan and coworkers [8] indicate that it is unlikely that the positive nitrogen interacts with Glu17. A crystallographic structure of TpR-TSST will certainly settle this point. Ser469' OG interacts with Glu-I carboxylate. This results in a global electro-neutral region in TpR that probably favors TSST binding (Figure 7) and destabilizes the binding of negatively charged substrates as GSSG. The fourth carboxylate group, common to all substrates, binds close to Lys67 in GR. This interaction is preserved in the TpR model.

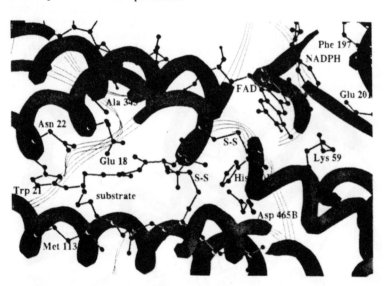

Figure 7- TSST docked in TpR active site.

TSST docked in the active site cavity of GR, at the place described above for GSSG and without moving side chains, creates large steric hindrance. Moreover, even if we allow for side chain relaxation, the positively charged nitrogen atom of spermidine must bind in a highly positively charged region in GR. This is a rather unfavorable situation. Thus, the specificity of GR for GSSG may be partially due to this electrostatic factor. Now, GSSG docked at the active site of TpR cannot be stabilized by the interactions set up by its four terminal carboxylate groups with the array of positively charged residues, which are missing in TpR. Instead, Glu17 now substitutes Ala34, thereby introducing an extra destabilizing factor. These results do not contradict site-directed mutagenesis studies [41], in which the substrate specificity of *E.coli* GR has been switched from GSSG to TSST by introducing the appropriate residues of TpR in the active site of GR

6. DOCKING OF INHIBITORS TO GR AND TpR

6.1. X-ray binding site of GSSG-retro in GR

The crystallographic structure at 2.4 A resolution of the retro-analogue of GSSG, namely N4-(malonyl-D-cysteinyl)-L-2,4-diaminobutyrate disulfide (GSSG-retro) in complex with GR has been recently deposited in the PDB [37]. Here, the peptide bond is reversed with respect to GSSG, giving rise to a different chirality. This structure is used to test the conclusions we have achieved from docking studies in GR and TpR of the natural substrates.

The binding site of GSSG-retro in GR is close to glutathione binding site (Figure 8). It is interesting to notice that the highest density of the complex between GR and GSSG-retro corresponds to the zwitterionic head of Glu-II and that the carboxyl group of Glu-I is the only other group of the inhibitor which may be superimposed to the corresponding one of the natural substrate. There is a 5 A shift of the disulfide bridge of GSSG-retro away from the protein bridge, which may account for the very low catalytic rate of this inhibitor.

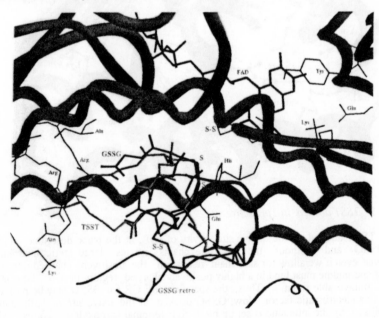

Figure 8- Comparison of the X-ray structure of GSSG-retro bonded to GR [37] and the docked models of GSSG and TSST in that same enzyme.

The X-ray study indicates that the strong binding contributions of Glu-I and Glu-II are preserved. One may guess at the importance of these interactions set up with the corresponding protein residues (Arg347 and Glu472', respectively) in the binding process of substrates or inhibitors to the GSSG binding site of GR.

Note that the malonyl moiety is intercalated between the S-S bridge of the two molecules.

6.2. Docking studies of nitrofuran inhibitors to GR and TpR.

Three different sites in GR and TpR were examined as possible binding sites for nitrofuran inhibitors of GR: the GSSG binding site, the NADPH binding site and the interface binding site.

6.2.1. Docking of nitrofurans at the NADPH binding site.

Before Tyr197/GR (or Phe197/TpR) closes the active site, there is plenty of space for a small molecule as NF-pyrazole to bind at the nicotinamide site of NADPH in GR and TpR: at least four different minimal energy structures were obtained in this case, which only differ in the relative orientation of the furane ring with respect to FAD. A priori, it is difficult to determine which of these different possibilities is the most likely. The calculation of differences in free energy of binding and quantum mechanical studies in progress in our groups will help clarify this aspect.

The main conclusion at the present preliminary stage is that a nitrofuran derivative may quite well occupy the NADPH binding site of the enzymes, with a lower affinity than the co-factor. If the nitrofuran derivatives bind at the NADPH binding site, then they should act as competitive inhibitors of $NADP^+$ in the reverse reaction. This conclusion is in accordance with the kinetics of the reverse reaction of nitrofurans in GR [42]. Moreover, it might also explain the observed reduction of nitrofurans in LipDH (an enzyme which catalyses the reverse reaction). However, this site does not explain the uncompetitive observed inhibition of the direct reaction.

6.2.2. Docking of nitrofurans at the interface binding site.

The interface binding site of quinones in GR is located on the two-fold symmetry axis of the dimer [31] and almost 15 A away from FAD. Quite interestingly, an asymmetric molecule may bind at this symmetry center in a completely different way with respect to the other two possible binding site: here, there will be two monomers bound to only one inhibitor molecule, while in the other cases there will be one inhibitor molecule bound per monomer. In spite of this fact, the occupancy factor of this site is twice the value of the other two possibilities for menadione and safranine.

The docking of NF-pyrazole at this site in GR was examined [43]. The H-bonds that may be established between the oxygen atoms of the inhibitor nitro group and His 75/ His75' of GR may act as a stabilizing factor for binding at this site. The furfuryliden ring of NF-pyrazole is stacked between Phe78/Phe78' rings of GR. This give rise to important hydrophobic interactions which, in this case, may also favor binding. The binding of nitrofuran derivatives to the regulatory site of TpR is more difficult to occur. This is due to extensive mutations of the enzyme in this region relatively to GR. This site in TpR is highly polar: His75/His75' and Phe78/Phe78' of GR have been exchanged respectively to Tyr75/Tyr75' and Gln78/Gln78' in this case. Also, the distance between the only two aromatic rings in TpR pocket is too big to account for hydrophobic stabilization in a stacking of the furfuryliden ring of NF-pyrazole. The same situation is observed for LipDH.

Clearly, this site differentiates GR from TpR: GR may be inhibited by the binding of an inhibitor at this position, while this possibility is unlikely to occur in TpR.

6.2.3. Docking of nitrofurans at the substrate active site

Two different regions were detected in the substrate binding zone of GR which are suited to bind the nitro moiety of nitrofurans [36]. There is a very flexible zone in the GS-I binding region: the groups of Arg 37, 38, 347 and 478' are inside a sphere of 4 A radius, which creates a positively charged region in the outer part of the active site cleft. The carboxylate group of Gly-I and the α- carboxylates of Glu-I and Gly-II of GSSG bind in this region. Large negative molecules could also bind in this zone favored by coulombic forces and simultaneously establish additional bonds in other regions of the enzyme to attain low binding entropy. Thus, nitrofurans can bind to GR in this region in a way similar to the carboxylate in Gly-I or in GSSG-retro Gly-I. This zone is instead a polar region, with charge -1, in TpR from C. fasciculata, T. congolense and T. cruzi and contains Glu-34, which is an Ala in GR. The difference in net charge in GR and TpR at this zone, that binds the spermidine moiety of TSST, should prevent the binding of the negatively charged nitro group to TpR and is an interesting target for designing specific inhibitors for this enzyme. However, this fact is in apparent contradiction with the low observed inhibition of TpR by nitrofuran derivatives [44].

There is another place in the active site binding region suited for binding nitrofuran derivatives in both GR and TpR. This is a deep pocket between both monomers where nitrofurans can bind very tightly, mimicking the binding of Glu-II, with the nitro groups occupying the same position as the carboxylate group of the substrate. In this position, the nitro group is probably bonded to Lys67 through a water molecule and hydrogen bonded to residues 407' and 406. The pirazoyl moiety of the nitrofuran derivative makes van de Waals contacts with the disulfide bridge. This site is very conserved in all the enzymes of the family except LipDH. Here, the amino acids involved in binding are conserved, but Lys 67 occupies the place of the missing Lys66, forming a salt bridge with Glu201.

7. CORRELATIONS BETWEEN THE INHIBITOR BINDING SITE AND THE CATALYTIC MECHANISM OF THE ENZYME: CONJECTURES.

The ping-pong catalytic mechanism of GR and TpR is envisaged in the literature as a two steps process [21]. In the first one, the enzyme is reduced to a stable form and NADPH is oxidized to $NADP^+$, which then leaves its binding pocket. In the sequence, the disulfide bridge of the enzyme is opened and interacts with that of the substrate, forming a mixed disulfide, while, according to our preliminary quantum chemical calculations, it is closed but charged, with subsequent release of GS-II. Thus, electrons flow from NADPH to the substrate GSSG via flavin and the redoxactive protein disulfide bridge. The forward reaction is inhibited by nitrofurans and biochemical data show that they do not compete either for the NADPH or the GSSG binding sites in the oxidized enzyme [45]. The effect of these nitro compounds on the reverse reaction of GR [42] indicates that the mechanism of this reaction is different from that of the forward one. However, the kinetic data do not discard a competition for the substrate binding site in this case.

The binding of nitrofurans at the NADPH site in all the enzymes, as studied here, should result in a competitive mechanism, which is not observed experimentally. On the other side, the binding of these nitro-derivatives at the regulative site of GR could account for the observed non-competitive mechanism. This hypothesis is in accordance with experimental evidences showing that TpR is weakly inhibited by nitrofurans, which agree with our results, as we found that it is rather unlikely that a nitrofuran should bind there. However, this site is 15 A apart from the active site. When binding to the regulative site of GR, the nitro-compounds could influence the redox-active disulfide via electron transfer polarization using as a bridge the α–helix from residues 63 to 80, which contains one of the active site Cys residues and those involved in binding at this site (His 75/His75' and Phe78/Phe78'). Also, they could influence FAD via H-bonding to Lys66 in the same helix. This possibility will be elicited by theoretical calculations.

The above analysis, although conjectural, lead us to suggest two binding sites as the most suited for nitrofuran both in GR and TpR: the NADPH binding site and the substrate binding site. Depending on the relative concentrations of inhibitor and natural substrate, one or the other site should be occupied. The substrate binding site might be of higher affinity for nitrofurans and may be preferentially occupied by it when the inhibitor concentration is relatively low. The NADPH site should be of low affinity for nitrofurans and NADPH should occupy it preferentially to the inhibitor. However, in the reverse reaction, the affinity of $NADP^+$ should differ to that of NADPH and be lower than that of the nitrofuran. Also, the concentration of the latter is higher than that of $NADP^+$. Thus, the nitrofuran may occupy this site and be reduced by the enzyme in the reverse reaction, in spite of the lack of significative inhibition at this site in the forward reaction. These conjectures have to be experimentally tested before some reasonable level of credibility can be assigned.

The description so far obtained have helped improve our understanding of the interactions between the enzymes and ligands such as the natural substrates and some known inhibitors that might be relevant for drug design. This information may give valuable hints to design selective inhibitors for TpR. Of course, the analysis is still qualitative in nature. Further quantitative free energy difference calculations would be required to include important entropic factors.

REFERENCES

1. Marr, J.J. & Docampo, R., *Rev. Infect. Dis.*, 8 (1986) 884.
2. van Voorhis, W.C., *Drugs, 40* (1990) 176.
3. Nagel, R., *Mutation Res., 191* (1987) 17.
4. Korolkovas, A., *Essentials of Medicinal Chemistry*, Wiley-Interscience, New York (1988).
5. Henderson, G., Ulrich, P., Fairlamb, A.H., Rosenberg, I., Pereira, M., Selas, M. & Cerami, A.,*Proc. Natl. Acad. Sci USA, 85* (1988) 5374.
6. Henderson, G.B., Murgolo, N.J., Kuryian, J., Osapay, K., Kominos, D., Berry, A., Scrutton, N.S., Hinchliffe, N.W., Perham, R.N. & Cerami, A., *Proc. Natl. Acad. Sci. USA, 88* (1991) 8769.
7. Horjales, E., Oliva, B., Stamato, F.M.L.G., Paulino-Blumenfeld, M., Nilsson, O. & Tapia, O., *Mol. Eng., 1* (1992) 357.
8. Kuryian, J., Kong, X.P., Krisna, T.S.R., Sweet, R.M., Murgolo, N.J., Field, H., Cerami, A. & Henderson, G.B., *Proc. Natl. Acad. Sci. USA, 88* (1991) 8764.
9. Hunter, W.N., Bailey, S., Habash, J., Harrop, S.J., Helliwell, J.R., Aboagye-Kwarteng, T., Smith, K. & Fairlamb, A.H., *J. Mol. Biol., 227* (1992) 322.

152

10. Fairlamb, A.H., *Parasitology, 99* (1989) S93.
11. Fairlamb, A.H., *Trans. Roy. Soc. Trop. Med. Hyg., 84* (1990) 613.
12. Castro, J.A. & Diaz de Toranzo, E.G., *Biom. Env. Sci., 1* (1988) 19.
13. Grinblat, L., Sreider, C. & Stoppani, A.O.M., *Biochem. Pharmacol., 38* (1989) 767.
14. Williams, C.H., in *The Enzymes*, Boyer, P.D. (ed.), Academic Press, New York (1976) 89.
15. Williams, C.H., in *Chemistry and Biochemistry of Flavoenzymes*, Muller, F. (ed.), CRC Press Inc., Boca Raton FLDA (1992) 121.
16. Worthington, D.J. & Rosemayer, M.A., *Eur. J. Biochem., 48* (1974) 167.
17. Greer, S. & Perthan, R.N., *Biochemistry, 25* (1986) 2736.
18. Westphal, A.H. & DeKok, A., *Eur. J. Biochem., 172* (1983) 299.
19. Shames, S.L., Kimmel, B.E., Peoples, O.P., Agabian, N. & Walsh, C.T., *Biochemistry, 27* (1988) 50.
20. Clark, A.R., Atkinson, T. & Holbrook, J.J., *Trends Biochem. Sci., 14* (1989) 145.
21. Pai, E. & Schulz, G.E., *J. Biol. Chem., 258* (1983) 1752.
22. Mattevi, A., Schierbeek, A.J. & Hol, W.G.J., *J. Mol. Biol., 220* (1991) 975.
23. Karplus, P.A. & Schulz, G.E., *J. Mol. Biol., 195* (1987) 701.
24. Jones, T.A., *J. Apppl. Cryst., 11* (1978) 268.
25. Cambillau, C. & Horjales, E., *J. Mol. Graph., 5* (1987) 174.
26. van Gunsteren, W.F. & Berendsen, H.J.C., Groningen Molecular Simulation (GROMOS) Library Manual, (ed.), BIOMOS B.V., Nijenborgh 16, Grorningen, The Netherlands (1987).
27. Aqvist, J., van Gunsteren, W.F., Leijonmarck, M. & Tapia, O., *J. Mol. Biol., 83* (1985) 461.
28. Rice, D.W., Schulz, G.E. & Guest, J.R., *J. Mol. Biol., 174* (1984) 483.
29. Karplus, P.A. & Schulz, G. E., *J. Mol. Biol., 210* (1989) 163.
30. Petsko, G.A., *Nature, 352* (1991) 104.
31. Karplus, P.A., Pai, E.FG. & Schulz, G.E., *Eur. J. Biochem., 178* (1989) 6124
32. Fairlamb, A.H., Blackburn, P., Ulrich, P., Chait, B.T. & Cerami, A., *Science, 227* (1985) 1485.
33. Allinger, N.L. & Lii, J.H., *J. Comp. Chem., 8* (1987) 1146.
34. Paulino-Blumenfeld, M., Hikichi, N., Hansz, M. & Ventura, O., *J. Mol. Struct., 210* (1990) 467.
35. Paulino-Blumenfeld, M., Hikichi, N., Hansz, M. & Stamato, F.M.L.G., V Simposio Brasileiro de Quimica Teorica, Caxambu, Brasil (1989).
36. Paulino-Blumenfeld, M. & Horjales, E., International Symposium on Crystallography and Molecular Biology, Guaruja, Brasil (1990).
37. Janes, W. & Schulz, G.E., *J. Biol. Chem. , 265* (1990) 10443.
38. Mester. B., Elguero, J., Claramunt, R.M., Castanys, S., Mascaro, M.L., Osuna, A., Villaplana, M.J. & Molina, P., *Arch. Pharm. (Weinheim), 320* (1987) 115.
39. Dubin, M., Fernandez-Villamil, H., Paulino-Blumenfeld, M. & Stoppani, A.O.M., *Free Rad. Res. Comm., 14* (1991) 419.
40. Sreider, C.M., Grinblat, L. & Stoppani, A.O.M., *Biochem. Pharmacol., 40* (1990) 1849.
41. Bradley, M., Bucheler, U.S. & Walsh, C.T., *Biochemistry, 30* (1991) 6124.
42. Rakauskiené, G.A., Cenas, N.K. & Kulys, J.J., *FEBS Lett., 243* (1989) 33.
43. Horjales, E., Stamato, F.M.L.G. & Ambrosio, M.I., *Mem. Inst. Oswaldo Cruz (Supl.) I, 86* (1991) 235.
44. Jockers-Scherubl, M.C., Schirmer, R.H. & Krauth-Siegel, L., *Eur. J. Biochem., 180* (1989) 267.
45. Cenas, N.K., Bironaité, D.A., Kulys, J.J & Sukhova, N.M., *Biochem. Biophys. Acta, 1073* (1991) 195.

The authors acknowledge the financial support from FAPESP and RHAE/Bio (Brazil) and SAREC (Sweden).

Permanent addresses:
E.H.: Facultad de Ciencias, Universidad de la Republica, Montevideo, Uruguay.
B.O.: Instituto de Biologia Fonamental, Universitat Autonoma de Barcelona, Spain.

Molecular Modeling at Corning, Inc.

Michael Teter

Engineering Fellow

The development of new inorganic materials at Corning, Inc. has had a long and distinguished history. A short and necessarily incomplete list of important materials discoveries at Corning includes low expansion borosilicate glasses such as Pyrex, ultra-low expansion glasses made by flame hydrolysis, silicones, fiberglass, the complex field of glass-ceramics which range from ultra-high strength materials to machinable mica based forms, photochromic glasses, x-ray absorbing television glasses, extrudable oriented ceramics, and the ultra-pure materials for optical fibers. While extremely successful, the development of new materials at Corning has been a difficult process. From the start of a research project until the manufacture of a commercial product can take from 5-10 years, and the vast majority of research projects fail. One important factor at the heart of the long development times and the high failure rate is a lack of fundamental understanding concerning reactions and structure in amorphous materials. New materials are developed by a mixture of intuition and empiricism.

From the time of Zachariason[1], we have had reasonable ideas about glass structure. In summary, first neighbor environments look almost exactly the same as they do in crystals, but these building blocks are put together in a continuous random network. Electro-neutrality is preserved in the smallest possible volume consistent with ion sizes. Unfortunately, the bonding differences between a glass such as fused silica and the crystal structure of quartz are reasonably subtle. We know the structure of quartz through the analysis of x-ray diffraction experiments. The repeating nature of the quartz crystal gives many diffraction peaks which allow the structure to be unambiguously extracted. Fused silica has no repeating units other than the SiO_4 tetrahedra and the strong diffraction peaks are related to these. The bonding differences between crystalline and amorphous silica occur primarily at third neighbors and beyond, and it is precisely this information which is washed out by diffraction from the amorphous structure. Most other probes again give strong information about first neighbors, weak results about second and are washed out by third neighbors. Thus we are not certain about the structure of the most prototypical of all glasses, fused silica, which serves as a backbone for most of the silicate glasses.

Doris Evans, an x-ray crystallographer at Corning, built a model of fused silica in the mid 1960's[2] to see if a model could be built with the constraint that the only construction variable was the bonding angle between SiO_4 tetrahedra. She built a model of over 1200 tetrahedra and was able to convince herself that the procedure could be continued indefinitely. A careful analysis of the model showed that it could nearly reproduce most of the known properties of fused silica. The most important discrepancy was in the density A careful study of variations in the density showed that it could be varied over 10% without rearranging the bonding or violating the constraint of perfect tetrahedra.[3] There had to be something else other than a geometric constraint which determined the final values of the physical properties. Clearly structures arrange themselves to minimize the free energy at elevated temperatures. At low temperatures, the product of temperature and entropy becomes negligible and the free energy converges to the potential energy. Fused silica is a child of the elevated temperatures at which its bonding arrangements are formed and the relaxation in geometry but not bonding at lower temperatures. Consequently it was felt that the only way to attack the problem of molecular structure of glasses was through molecular dynamical modeling at high temperatures which could properly account for the dynamic effects of entropy coupled with a slow decrease in temperature to relax the geometry. Born-Meyer-Huggins pair potentials were created for silicon ions and oxygen ions, and I performed molecular modeling for fused silica for over 1000 ions. This led to a model which resembled the model of Evans but was different in many of the details. She then asked me if my potentials could reproduce the structure of the crystalline polymorphs of silica i. e. quartz, cristobalite, coehsite, tridymite and stishovite. If I was claiming to be able to model SiO_2, I should be able to reproduce the structures that it found to be the most natural ways to arrange itself in. I found to my dismay that I could not reproduce such structures. This led to Evans Law of Glass Structure: "Accept no model of glass structure which cannot reproduce as stable structures the underlying crystalline polymorphs of similar composition."

Unwilling to give up on the effort invested in the molecular dynamics computer program I proceeded to model the structure of another prototypical glass, B_2O_3. The fundamental building unit in boric oxide is the BO_3 triangle. The basic issue of the structure of B_2O_3 glass is how these triangles are bonded together. The prevailing idea suggested by Goubeau and Keller[4], strengthened by the infra-red work of Krogh-Moe[5] and further developed by the x-ray diffraction analysis of Mozzi and Warren[6] is that the structure of boric oxide glass is primarily one in which 3 BO_3 triangles are bonded in a planar boroxol ring, with a $B - O - B$ bonding angle of 120 degrees.

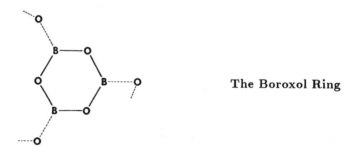

The Boroxol Ring

The molecular dynamics run showed that in a cluster of over 1000 ions, there were exactly two boroxol rings and that the average boron-boron bonding angle was over 150 degrees. Since the x-ray diffraction results and my structure seemed to agree reasonably well, although there was some concern since the x-ray scattering factor for boron was negligible, I was encouraged, and began to write the results for publication. Two things happened to stop this. First, Thomas Soules[7] of General Electric published his results which were nearly identical to mine and second Adrian Wright published the results of neutron diffraction in B_2O_3 glass[8] which showed that the $B - O - B$ bonding angle was unambiguously 120 degrees. Wright's conclusion after studing several models was in agreement with Warren and Mozzi that boroxol rings were the dominant structure. This was strengthened by the work of Windisch and Risen[9] who measured the Raman spectra of B_2O_3 glass and found by isotopic substitution that the single large peak at 808 cm^{-1} was a breathing mode of the boroxol ring.

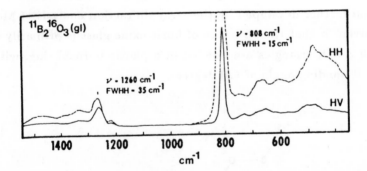

At Doris's suggestion, I performed the calculation on the boric-oxide crystal structure and could not reproduce it as a stable structure with my potentials. At this point I gave up treating oxides as simple ionic structures, and retreated from molecular dynamics for nearly a decade.

In the early 1980's, the need for a first principles treatment of glass chemistry was becoming more urgent, and I undertook a new survey of the field to find changes. At MIT, talks with Professor John Joannopoulos convinced me that the state of the art of density functional quantum mechanics developed by Kohn and Sham[10] had reached the point where solids could be realistically modeled without parameterization.

The typical condensed matter calculation modeled a semiconductor with a few atoms in a periodic cell. Planewaves and pseudopotentials were used, and to model silicon required approximately 50-100 planewaves per atom. The diagonalization of matrices a few hundred square was not a demanding task, but since diagonalization grew as the number of planewaves cubed, much bigger problems were computationally intractible. Because of the deep nonlocal pseudopotentials required to accurately model oxygen, oxygen requires ten times as many planewaves per atom as silicon. To simulate quartz would require the diagonalization of 10000 by 10000 matrices. In 1985, Car and Parrinello[11] wrote a paper on quantum mechanical molecular dynamics which emphasized the iterative approach to matrix diagonalization and showed how to derive an eigenvector by a method which scaled nearly linearly with the number of planewaves instead of cubically. The new method required local pseudopotentials which would not work for oxygen and was still too slow. Allen and Teter[12] showed how oxides could be modeled successfully, and Teter, Payne and Allen[13] showed how eigenvectors could be extracted most efficiently. The details of the resulting iterative diagonalization along with a review of the planewave pseudopotential method may be found in the review article by Payne et al.[14] For the first time, this made the planewave pseudopotential treatment of oxides viable. Five years ago, we began the systematic treatment of silicates using *ab initio* methods and have fulfilled Doris Evans' challenge that all crystalline polymorphs of silica be faithfully modeled by this method. The tables below show the calculated versus measured structural parameters for several silica and titania crystals.[15]

Table I. α-Quartz Structure (Trigonal $P3_221$)

Parameter	Experiment*	Theory	Δ (%)
a (nm)	0.49134	0.4880	−0.68
c (nm)	0.54052	0.5370	−0.65
Si (x)	0.46987(9)	0.4638	−1.3
O (x)	0.4141(2)	0.4081	−1.4
O (y)	0.2681(2)	0.2758	+2.9
O (z)	0.2855(1)	0.2782	−2.6

Table II. α-Cristobalite Structure (Tetragonal $P4_12_12$)

Parameter	Experiment*	Theory†	Δ (%)
a (nm)	0.49570(1)	0.4959	0.04
c (nm)	0.68903(2)	0.6906	0.23
Si $(x = y)$	0.3047(2)	0.3030	−0.56
O (x)	0.2381(2)	0.2380	−0.04
O (y)	0.1109(2)	0.1112	0.27
O (z)	0.1826(1)	0.1825	−0.05

Table III. Stishovite Structure (Tetragonal $P4_2/mnm$)

Parameter	Experiment*	Theory	Δ (%)
a (nm)	0.41773(1)	0.4255	1.9
c (nm)	0.26655(1)	0.2604	−2.3
O $(x = y)$	0.30614(3)	0.3082	0.67

Table IV. Rutile (TiO_2) Structure (Tetragonal $P4_2/mnm$)

Parameter	Experiment*	Theory	Δ (%)
a (nm)	0.4594	0.4584	−0.22
c (nm)	0.2958	0.2961	0.10
O $(x = y)$	0.305	0.3044	−0.20

From these studies a simple method of parameterizing interactions between the electrons and ions was determined, and studies were made of the structures of silica and boric oxide which included quantum effects.

The results of the silica simulation were minor changes in the structure I had derived years earlier. When the boric oxide structure was modeled, however, there were major changes in the results. This time the modeled pair distribution function exactly matched that determined experimentally by both x-ray and neutron diffraction. This meant that the $B - O - B$ bonding angle was now correct at 120 degrees. The original density had been too small by over 5% and now was correct to better than 1%. A study of the bonding topology led to the most significant surprise. There was again only a few boroxol rings out of the several thousand ions modeled. Because, however, the $O - B - O$ and the $B - O - B$ bonding angles were now both 120 degrees, the radial distributions of this model and one containing a large fraction of boroxol rings with the same bonding angles were very similar.

That this result is reasonable may be easily seen in the figure below.

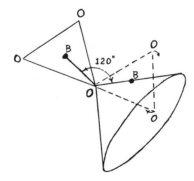

Although the bonding angle of the two triangles must be 120 degrees, the entire cone angle is available. when that is fixed, the entire dihedral angle is still available. The probability that three triangles would arrange themselves in a planar structure to give the boroxol ring must be very small unless there is some energetic advantage to the arrangement. Quantum mechanical simulation shows no such energetic advantage since the structure is not a resonance one.

Only one problem remained. How could one explain the Raman spectrum whose only major peak had been tied definitively to a breathing mode of the boroxol ring. There are two possible causes for a large peak in spectra. The first is a large density of states. The second is a large transition probability. There is an independent measure of vibrational density of states in a solid, inelastic neutron scattering. Roger Sinclair[16] ran such a measurement and found practically zero density of states at the position of the strong Raman peak.

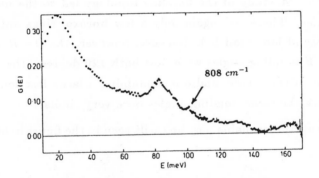

This means that the strong peak is due solely to an extremely large transition probability.

From another point of view, the optical phonons in quartz were simulated using the ab initio codes.[17] The results were with a few percent of experiment, but more importantly, the interaction between ions could be quantitatively be measured. A study of the Born Effective-Charge Tensor shows that no model which treats oxygen ions as charged balls can effectively model the dynamics in oxides. The polarization of the ions is at least as important as their charge in determining the correct interactions.

In summary, we began simulations with methods which from several points of view were shown to be wrong. By making advances in the calculation algorithms for solutions to Schroedinger's equation, we have been able to bring quantum mechanical methods to glass chemistry. We now have a simple way to summarize these results and are able, for the first time, to derive structures for simple glasses. For even these cases, the results are sometimes surprising as in the case of the boroxol rings in boric oxide. The next goal is to determine the physical causes of some of the more important effects in glass and ultimately to generate a new material which owes its existance to insights gained through simulation.

References

1. W. H. Zachariason, J. Am. Chem. Soc. 54 (1932) 3841.

2. D. L. Evans and S. V. King, Nature 212 (1966) 1353.

3. D. L. Evans and M. P. Teter, The Structure of Non-Crystalline Materials, Edited by P. H. Gaskell, Taylor and Francis London (1977) 53.

4. J. Goubeau and H. Keller, Z. Anorg. chem. 272 (1953) 303.

5. J. Kroh-Moe, J. Non-Crystalline Solids 1 (1969) 269.

6. R. L. Mozzi and B. E. Warren, J. Appl. Crystallogr. 3 (1970) 251.

7. T. F. Soules, J. Chem. Phys. 73 (1980) 4032.

8. P. A. V. Johnson, A. C. Wright and R. N. Sinclair, J. Non-Crystalline Solids 50 (1982) 281.

9. C. F. Windisch, Jr. and W. M. Reisen, Jr., J. Non-Crystalline Solids 48 (1982) 307.

10. W. Kohn and L. J. Sham, Phys. Rev A 140 (1965) 1133.

11. R. Car and M. Parrinello, Phys. Rev. Lett. 55 (1985) 2471.

12. D. C. Allan and M. P. Teter, Phys. Rev. Lett. 59 (1987) 1136.

13. M. P. Teter, M. C. Payne and D. C. Allan, Phys. Rev. B 40 (1989) 12255.

14. M. C. Payne, M. P. Teter, D. C. Allan, T. A. Arias and J. D. Joannopoulos, Rev. Mod. Phys. 64 (1992) 1045.

15. D. C. Allan and M. P. Teter, J. Am Ceram. Soc. 73 (1990) 3309.

16. R. N. Sinclair, J. Non-Crystalline Solids 76 (1985) 61.

17. X. Gonze, D. C. Allan and M. P. Teter, Phys. Rev. Lett 68 (1992) 3603.